Qualitative and Quantitative Analysis of Scientific and Scholarly Communication

Series editors

Wolfgang Glänzel, Katholieke Univeristeit Leuven, Leuven, Belgium
Andras Schubert, Hungarian Academy of Sciences, Budapest, Hungary

More information about this series at http://www.springer.com/series/13902

Yuehong (Helen) Zhang

Against Plagiarism

A Guide for Editors and Authors

Springer

Yuehong (Helen) Zhang
Zhejiang University Press
Hangzhou
China

This book is part of a research study commissioned by the Committee on Publication Ethics (COPE) with the aim of developing evidence-based guidance for authors and editors on how to avoid different kinds of plagiarism (http://publicationethics.org/resources/research); the author also acknowledges the support of the Project for Enhancing International Impact of China STM Journal (http://www.cast.org.cn/n35081/n35488/15215969.html) and the National Natural Science Foundation of China (No. 30824802).

ISSN 2365-8371 ISSN 2365-838X (electronic)
Qualitative and Quantitative Analysis of Scientific and Scholarly Communication
ISBN 978-3-319-24158-6 ISBN 978-3-319-24160-9 (eBook)
DOI 10.1007/978-3-319-24160-9

Library of Congress Control Number: 2015950004

Springer Cham Heidelberg New York Dordrecht London

Printed on acid-free paper

Springer International Publishing AG Switzerland is part of Springer Science+Business Media (www.springer.com)

Foreword

Pretending that someone else's work is one's own, i.e., using it without attribution, is plagiarism; the plagiarized item may be a section of text, a figure, a table, or even an idea. And re-using sections of text, figures, tables, or data from one's own previously published work without crediting the original publication is also plagiarism, in this case self-plagiarism. Plagiarism can occur in all kinds of creative work—writing, music, film, etc.—but this book is specifically concerned with plagiarism in scholarly journal articles.

Plagiarism is thus a form of theft (stealing someone else's work) and dishonesty (passing it off as one's own). If the copied material is extensive, it may involve copyright theft as well; there are no set rules on how much material may be quoted before copyright permission is required. What is more, it wastes the precious time of editors, reviewers, and readers, obliging them to read something that is not new, under the mistaken impression that it is original.

Plagiarism is not a new problem: the Roman poet Martial, who lived in the first century CE, complained that other poets were copying and issuing his works without attribution [1]. However, the Internet makes it much simpler to plagiarize—an author can easily copy passages from other works and paste them into her own. The problem occurs at all levels of authorship; there have been instances of very well-respected authors being found out in plagiarism [2]. However, it does seem to be more common among inexperienced authors—who may be unaware of the correct standards of conduct for academic authorship—and in authors from cultures where repetition of a master's words is considered to be highly respectful [3–5].

While the Internet makes plagiarism easier to commit, it also makes it much easier to detect—tools such as iThenticate (the software used by CrossCheck) enable journal editors and publishers to check whether significant strings of text in a submitted article are similar or identical to those in other published works already available within the tool's database [6].

There are pitfalls in relying too heavily on such tools, however. Not all relevant published work will necessarily be in the database used by the tool of choice; this is where the expertise of journal editors and peer reviewers is invaluable, since they should be very familiar with the literature in their field and thus able to recognize material taken from other sources. The detection of similar or identical strings of text does not always indicate plagiarism; the re-used material may in fact have been credited to its source, or it may represent a standard description of something, such as a common procedure. The editor must always look at the highlighted sections of text and make a decision as to whether or not plagiarism is involved. And at the same time, tools such as these cannot identify all instances of plagiarism, such as re-used illustrative material, or—of course—ideas. The Committee on Publication Ethics, COPE, has very useful guidance on its Web site (http://publicationethics.org/).

Every journal should have a clear written policy on plagiarism (and other ethical problems) for the guidance not only of authors, but also of editors and reviewers. The policy should be based on the best available guidance and examples, such as the Committee on Publication Ethics' own guidelines, the Office of Research Integrity's 'Guide to Ethical Writing' (http://ori.hhs.gov/avoiding-plagiarism-self-plagiarism-and-other-questionable-writing-practices-guide-ethical-writing), and examples from the most respected journals in the field. The journal should set high standards and should make clear what will happen if those standards are not met.

Authors can avoid suspicions of plagiarism by always identifying clearly any material that has been taken from another source (whether the author's own work or someone else's). Short quoted extracts should be indicated in quotation marks, while longer ones should be shown as indented paragraphs. In either case, a full citation must be given to the original source; if the extract is at all extensive, copyright permission must also be sought from the publisher as well as the author, since the latter may not always be the copyright holder. Changing a few words or even paraphrasing someone else's words does not alter the situation, since the content is still not original—the original author and source should still be credited.

In this book, Yuehong (Helen) Zhang—who has rapidly become an acknowledged expert on plagiarism, particularly in less developed countries—draws together a number of research articles reporting studies carried out, with her colleagues, on various aspects of plagiarism. These examine the attitudes and experiences of authors, reviewers, and editors in a number of different disciplines. Building on this knowledge base, and drawing on the best available international guidance, Zhang puts forward sound practical advice for both authors and journal editors on how to avoid committing plagiarism, how to find it, and what to do when it is found. If readers of this book follow the guidelines it provides, the incidence of plagiarism—both unintentional and intentional—will be considerably reduced, which will greatly benefit the integrity of scientific research publication.

Sally Morris

Sally Morris has worked in journal and book publishing for over forty years, for presses including Oxford University Press, Churchill Livingstone and Wiley, and as CEO of the Association of Learned and Professional Society Publishers and editor-in-chief of *Learned Publishing*. She has played a leading role in many industry and publisher/library groups and has written and lectured widely on copyright and journal publishing.

References

1. Martial, *Epigrams* 1.53.
2. Cadbury, D. 2010. *The Dinosaur hunters: a true story of scientific rivalry and the discovery of the prehistoric world*. London: Fourth Estate.
3. Qiu, J. 2010. Publish or perish in China. *Nature* 463: 142–43. doi:10.1038/463142a.
4. Fang, F. C., F. G. Steen, and A. Casadevalli. 2012. Misconduct accounts for the majority of retracted scientific publications. *Proceedings of the National Academy of Sciences* 109: 17028–33. doi:10.1073/pnas.1212247109.
5. Thatcher, S. G. China's copyright dilemma. *Learned Publishing* 21: 278084. doi:10.1087/095315108x356707.
6. Jungbauer, B. 2010. CrossCheck—Plagiarism police on the scene. *Science 2.0*. July 7, 2010. http://www.science20.com/truth_universally_acknowledged/crosscheck_plagiarism_police_scene.

Preface: Why I Wrote This Book

Back on the morning of September 8, 2010, a normal working day as a journal editor, when I went to open my office e-mail in-box, the title of one particular e-mail—'Your Correspondence in *Nature*'—suddenly made my heart beat faster. I soon realized that the title of my recent short paper in *Nature* [1]—'Chinese journal finds 31 % of submissions plagiarized'—might be about to cause trouble. Comments about the paper rapidly accumulated both on *Nature*'s Web site [2] and on China's ScienceNet [3], and later that day my university principal, Yang, phoned me to ask why I had given my paper such a contentious title rather than the earlier title of 'Policing plagiarism in China is helped by innovative software' which he had seen in proof a few days earlier and indeed cited in his presentation at the Shanghai Scientific Journal Development Conference (also, as it happens, attended by *Nature*'s chief editor, Philip Campbell) the previous day. The dramatic title change was suggested at the last minute by a senior editor of *Nature*, who felt that 'the new title would have more impact and encourage more people to read the letter.' The outcome was that my normally quiet life was suddenly disrupted by many telephone calls and interviews from both domestic and foreign media; there was also a flurry of online comments [2, 3], which were a mixture of criticism, doubt, support, and understanding.

In those few days, I also received many letters from home and abroad, including one from Professor John Suppe, an eminent geoscientist and a member of the US National Academy of Sciences. The correspondence, and his letter in particular, opened my eyes and made me realize that I was doing the right thing, even though it had got me into trouble. Here is what his letter said:

> Dear Helen,
> Thank you for writing to me and congratulations for your publication in *Nature*, even if it is causing you some problems ... I still think you should be supported in China at the highest level and congratulated for this contribution. In fact the controversy should ultimately be a good thing for developing higher scientific success in China.
>
> ...

> I guess part of the problem is that there are a number of somewhat different unacceptable and dubious practices that are combined together and called plagiarism in the title. In your 2010 *Learned Publishing* paper [4] you explain very clearly what these are. Some are much more serious than others. For some types I think there may be legitimate disagreement, particularly in some specific cases. You stated in the abstract of the *Learned Publishing* article that it is important for the community to reach a consensus on these issues. I agree. The scientific and publishing community in China must reach consensus together in a way that promotes high quality in Chinese science.
>
> … It is very important that Chinese science and scholarly publications truly rise to the highest levels, just as China is aspiring to the highest levels in every sphere, such as sports and economic development. Clearly, cheating in sports would be viewed as unacceptable in the international community, even though people and groups in every part of the world attempt to cheat. It is the role, for example, of the Olympics Committee or other such bodies to apply strict standards of enforcement, even though it can lead to big public debate and controversy. In the same way, I am sure that the Chinese National Natural Science Foundation realizes that it must fully support the application of the highest standards of scientific excellence and excellence in standards of publication. The rewards come from true high quality original contributions to knowledge, not from taking shortcuts.
>
> However, there is a temptation sometimes to make a shortcut … I remember, many years ago there were major problems in Taiwan because of piracy of books, music, software, and other industrial products. It was finally realized that this piracy had to stop—it was not good for economic development in Taiwan. I guess the local authorities worked hard to enforce regulations and this was important for the development of a strong and innovative technical industry, competitive in the world.
>
> The same is true for scientific development. Plagiarism ultimately weakens the quality of the science and is very dangerous. It's a form of corruption. It is well documented that there is an inverse relationship between economic development and corruption. The lack of economic development of the Philippines over the last 50 years is widely ascribed to corruption. The rather weak performance of Italy both economically and scientifically is widely ascribed to corruption, even by the Italians. But it is a matter of degree, Italy does have some very great scientists, but the nation would be world class, at the level of France, Britain, Germany or Switzerland, if it were not for corruption.
>
> So I think it is important to China that some very strong people in China support you in this. It's a very important issue for the success of Chinese science. You might remember I wrote a short article on the growth of science for the 100th anniversary of Nanjing University [5] in which I forecast that the biggest contributor to growth of science worldwide in the 21st century would be China. However, it is equally clear that there are various things that can keep China from tasting the highest success. Plagiarism and other short cuts to a false success can easily weaken Chinese science. I'm sure that the leaders of science in China fully realize this and will give close attention.

The controversy was also reported by both Chinese and foreign media including CCTV-24 (China Central Television), *The New York Times* and National Public Radio (NPR). Here is a flavor of these reports:

> Last month a collection of scientific journals published by Zhejiang University in Hangzhou reignited the firestorm by publicizing results from a 20-month experiment with software that detects plagiarism. The software, called CrossCheck, rejected nearly a third of all submissions on suspicion that the content was pirated from previously published

> research… The journals, which specialize in medicine, physics, engineering and computer science, were the first in China to use the software…
>
> The journals' editor, Zhang Yuehong (Helen), emphasized that not all the flawed papers originated in China… Some were from South Korea, India and Iran… [6].
>
> For a decade, Helen Zhang has had a dream: to run an international scientific journal that meets international standards… In 2008, when her scientific publication, the *Journal of Zhejiang University-Science,* became the first in China to use CrossCheck text analysis software to spot plagiarism, Zhang was pleased to be a trailblazer. But when the first set of results came in, she was upset and horrified. 'In almost two years, we find about 31 percent of papers with unreasonable copying and plagiarism,' she says, shaking her head. 'This is true.' …When Zhang published these findings, she was criticized for bringing shame on Chinese scientists, even though she had emphasized that many of the papers were from overseas. China is forecast to become the world's leading innovator this year, overtaking the United States and Japan in the number of patent filings, according to Thomson Reuters. More scientific papers come out of China than out of any other country but the U.S., and Chinese leaders vow it will be a research superpower by 2020 [7].

The uproar caused by this topic in academic and publishing circles all over the world made me think about when it was that we, as academics and publishers, first began to pay close attention to the issue of 'plagiarism.' In my own experience as a scientific journal editor for more 30 years, I have noticed that in the last decade there have been more papers concerned with the topic of plagiarism, possibly because since the 1990s, due to the rise of the Internet, duplication has become easier. The proliferation of articles on this topic has increased since the arrival of tools for detecting plagiarism, such as iThenticate [8] (launched in 2004 and used by the CrossCheck service, which received the Association of Learned and Professional Society Publishers Award for Publishing Innovation in 2008) [9]. The Committee on Publication Ethics (COPE) [10], established in 1997, has focused on a wide range of ethical issues, including plagiarism in all its forms, and in 2008, COPE set up a Research Grant to help editors and publishers to study all aspects of publication ethics and, in particular, how to handle cases of research and publication misconduct.

We trialled CrossCheck with our own journal in 2008, but encountered various problems. This, combined with the above-mentioned background, encouraged us to apply for a COPE Research Grant, which we were fortunate enough to receive in early 2011 (Fig. 1) [11].

Our COPE-funded program, entitled 'CrossCheck guidance: an analysis of typical cases of plagiarism in different disciplines,' was duly carried out in 2011–2012. In our initial investigation, we focused primarily on three important questions:

1. What are journal publishers' and editors' attitudes to, and tolerance of, typical plagiarism in different disciplines?
2. What are the mainstream views and differences between editors in Western countries and non-Western countries?
3. How do journal publishers and editors worldwide use CrossCheck/iThenticate and how do they handle the statistics that it produces?

THE SCOOP FROM COPE

Y. H. (Helen) Zhang (right) and her research group (above)

COPE grant awarded for first time to recipients from China

COPE's December 2010 research grant was awarded to Yuehong (Helen) Zhang and Xiaoyan Jia of Zhejiang University in Hangzhou, China, for the project "CrossCheck Guidance: An Analysis of Typical Cases of Plagiarism in Different Disciplines."

Zhang, who is Journal Director in the Journals Department of Zhejiang University Press, and Jia, who is Editor of JZUS-A/B/C, the Journal of Zhejiang University-SCIENCE A/B/C (Applied Physics & Engineering, Biomedicine & Biotechnology, Computers & Electronics), with their editorial group, have already used the CrossCheck plagiarism detection software to identify potential cases of plagiarism in more than 2000 manuscripts submitted to their multidisciplinary journals in the past year. Zhang reported in *Nature* that 692 of 2,233 submissions contained unoriginal material (Zhang Y. *Nature* 2010;467:153).

With the COPE grant, Zhang and her group will select 3–5 representative cases of plagiarism in each discipline, make a detailed characterization of plagiarism in these typical cases, and provide suggestions for dealing with similar cases of plagiarism based on input from other CrossCheck users.

They plan to compile a handbook listing typical cases for CrossCheck users and authors worldwide. With this handbook, editors "can learn how to deal with different kinds of plagiarism in different disciplines when using CrossCheck" and authors "can learn more about plagiarism and CrossCheck, and how to avoid being accused of plagiarism."

COPE considers applications for grants twice a year, in June and December. The grants of up to £5000 are awarded to COPE members for a research project in publication ethics (see www.publicationethics.org/ research). The first grant was awarded in 2008, and one of the first year's grant recipients, Ana Marusic, will report on her research at the March 2011 COPE UK seminar (see page 2). 2010 marks the first time that the grant recipients are from China.

Zhang reported in *Nature* that 692 of 2,233 submissions contained unoriginal material

ETHICAL EDITING / SPRING 2011 C

Fig. 1 COPE Grant Award Report, *Ethical Editing*, Spring issue, 2011 (reproduced with the permission of COPE)

In order to answer these questions, and in an attempt to supplement the available ethical guidance for authors and editors in the scholarly publishing area, we conducted several worldwide surveys. To date, we have already published a few of

English language papers in international journals [12–18]. Our project is described on the COPE Web site [19] as follows:

> **CrossCheck guidance: an analysis of typical cases of plagiarism in different disciplines**
> Most plagiarism cannot be judged solely by the similarities discovered when using CrossCheck. Based on experience of cross checking more than 2000 manuscripts from approximately 50 countries in different parts of the world per year, this project aims to provide 3–5 typical cases of CrossChecked plagiarism in three different disciplines covered by the *Journal of Zhejiang University-SCIENCE A/B/C* (http://www.zju.edu.cn/jzus/) (*JZUS-A: Applied Physics and Engineering; JZUS-B: Biomedicine and Biotechnology; JZUS-C: Computers and Electronics*). The typical plagiarism case analysis will be made into a list or a handbook that will be classified by discipline. For editors, they can learn how to deal with different kinds of plagiarism in different disciplines when using CrossCheck. For authors, these lists can act as an instruction for authors on plagiarism, from which they can learn more about plagiarism and CrossCheck, and know how to avoid being accused of plagiarism.

The results of part of this research were also presented at the CrossRef 2011 Annual Meeting, USA, 15 November 2011 [20]. The purpose of this survey was to investigate the use of CrossCheck by journal editors in various different displines to detect plagiarism, and their attitudes to potential plagiarism, once discovered.

My aim in putting together this book is to gather all the findings from our various surveys of different disciplines in order to help journal editors, authors, and students from different subjects to learn more about plagiarism, including typical problems such as cut-and-paste, duplication of conference proceedings, self-plagiarism, team plagiarism, and review papers with a high level of similarity , and to find out how to detect plagiarism, how to deal with it, and how to avoid it.

Yuehong (Helen) Zhang

References

1. Zhang, Y. H. 2010. Correspondence: Chinese journal finds 31 % of submissions plagiarized. *Nature* 467(9): 153.
2. http://www.nature.com/nature/journal/v467/n7312/full/467153d.html.
3. http://news.sciencenet.cn/htmlnews/2010/9/237322.shtm.
4. Zhang, Y. H. 2010. CrossCheck: An effective tool for detecting plagiarism. *Learned Publishing* 23: 9–14.
5. Suppe, John. 2002. Exponential growth of geology, mathematics, and the physical sciences for the last two hundred years and prospects for the future. *Journal of Nanjing University-Natural Sciences, Special Issue for World Famous Scientists Forum* 38: 76–86.
6. Jacobs, A. 2010. *Rampant fraud threat to China's brisk ascent*. New York Times. www.nytimes.com/2010/10/07/world/asia/07fraud.html.
7. Lim, L. 2011. NPR: Plagiarism Plague Hinders China's Scientific Ambition. http://www.npr.org/2011/08/03/138937778/plagiarism-plague-hinders-chinas-scientific-ambition.
8. iThenticate. http://en.wikipedia.org/wiki/IThenticate.
9. ALPSP announces winners of 2008 Awards. www.alpsp.org.

10. COPE. http://publicationethics.org.
11. Wurz, J. 2011. COPE grant awarded for first time to recipients from China, *Ethical Editing* 3(1).
12. Zhang, Y. H, and X. Y. Jia. 2012. A survey on the use of CrossCheck for detecting plagiarism in journal articles. *Learned Publishing* 25(4): 292–07.
13. Zhang, Y. H., and I. McIntosh. 2012. How to stop plagiarism: Blacklist repeat offenders? *Nature* 481: 22.
14. Zhang, Y. H, X. Y. Jia, H. F. Lin, and X. F. Tan. 2013. Editorial: Be careful! Avoiding duplication: a case study. *Journal of Zhejiang University-SCIENCE B (Biomedicine & Biotechnology)* 14(4): 355–58.
15. Jia, X. Y., X. F. Tan, and Y. H. Zhang. 2014. Replication of the methods section in biosciences papers: is it plagiarism? *Scientometrics* 98: 337–45.
16. Zhang, Y. H., and X. Y. Jia. 2013. Republication of conference papers in journals? *Learned Publishing* 26(3): 189–96.
17. Zhang, X. X., Z. L. Huo, and Y. H. Zhang. 2014. Detecting and (not) dealing with plagiarism in an engineering paper: Beyond. CrossCheck—a case study. *Science and Engineering Ethics*.
18. Zhang, Y. H. 2013. Bilingual (multilingual) publications and duplicate publications: For or against? *Journal of Zhejiang University-SCIENCE A (Appl Phys & Eng)* 14(9): 687–90.
19. COPE Research Grant: December 2010, CrossCheck guidance: An analysis of typical cases of plagiarism in different disciplines. http://publicationethics.org/resources/research.
20. (Helen) Zhang, Y. H. 2011. A global survey on detecting plagiarism in journals using CrossCheck/Ithenticate, September 2011 in 2011 CrossRef. Annual Meeting (PDF 745kb).

Acknowledgments

I could not have completed the COPE research program and written this book without much substantial help not only from Sally Morris who provided active advice and professional guidance at all stages in the preparation of the whole project and all publications, as well as including Ian McIntosh, but also from my journal's editorial team: Jia Xiao-yan, Lin Han-feng, Zhang Xin-xin, Tan Xu-fei, Zhai Zi-yang, and Zhang Chun-jie. They are all editors of *Journal of Zhejiang University-SCIENCE A/B/C*, which are focused on different disciplines, each of which faces different plagiarism problems. Jia Xiao-yan, as my key assistant in the COPE project, not only analyzed a large amount of survey data but also—thanks to her life science PhD background—acted as co-author (with Tan Xu-fei et al.) of a survey on 'Replication of the methods section in biosciences papers: is it plagiarism?' Zhang Xin-xin, an editor of *JZUS-A*, looked in detail at detecting and (not) dealing with plagiarism in one particular engineering paper, with considerable assistance from Dr. Alex Fox. The support and suggestions of editors Lin Han-feng of *JZUS-B* and Zhai Zi-yang of *JZUS-C* are also much appreciated; in addition, thanks are due to PhD student Jiang Yi-sheng in the Department of Mathematics, Zhejiang University, for his statistical analysis. I have learned a great deal from all of you—thank you.

Contents

About the Author

Yuehong (Helen) Zhang has been as chief editor of *Journal of Zhejiang University-SCIENCE A/B & FITEE* and Journal Director of Zhejiang University Press for 16 years. She is also a council member of ALPSP (Association of Learned and Professional Society Publishers), board member of PILA/CrossRef, and vice president of Society of China University Journals, respectively. Helen is very active in STM journal publishing area. She has for many times being invited to give speeches at international conferences, including at the 4th World Conference on Research Integrity in Brazil, the 11th International Conference of Science Editors (in Beijing), and the 12th International Conference of Science Editors (in Mexico) etc. She has published several short papers in *Nature* and many papers in international journals. In 2010, she received a research fund from Committee on Publication Ethics (COPE). And she, with her team, already published many papers focused on CrossCheck guidance: an analysis of typical cases of plagiarism in different disciplines.

Part I
General Plagiarism Issues

Chapter 1
What Is Plagiarism?

Plagiarism is a form of academic dishonesty, and the importance of honesty in human society is not a new topic!

Famous phrases of Confucius, the Chinese philosopher of the 5th century BCE, include: 'Regulate the mind (Zhengxin)', 'Make the purpose sincere' (Chengyi) and 'Cultivate personal virtue (Xiushen)' [1]. And as the famous ancient Roman philosopher, statesman, lawyer, orator and political theorist Marcus Tullius Cicero wrote in the first century BCE [2], 'Where is there dignity unless there is honesty?' These principles apply just as strongly today. As Bosman said in his book, 'a person of integrity is honest, upright and devoid of duplicity, someone who displays consistency and strength of moral conviction, with a consequent resistance to acting against an internalised moral code' [3]. Thus, since ancient times, integrity has been a perennial topic of interest to human society.

However, communication technology has made a big difference; today's Internet makes it easier not only to commit plagiarism, but also to detect it. Now that we live in the network and digital age, it is no longer true that seeing is believing. 'Not so long ago, everyone knew that a photo doesn't lie. Today, image manipulation is not only possible but common. Likewise, the fact that research has been published doesn't mean that it's true' [4]. Editors of academic journals now have to spend a great deal of time dealing with a variety of forms of authorial misconduct, in particular plagiarism.

In recent years plagiarism has become a high-profile issue for academic journals; there have been many articles, books, and seminars discussing how to stop plagiarism in academic publications [5–15], which will be helpful documents for your references, especially for an example in 2012, *Nature* invited ten experts to explain how to stamp it out in its Comment [5].

Definitions of Plagiarism

Unlike the concept of copyright, which has had a legal basis for the past 300 years, the notion of plagiarism is perhaps more subjective; although it may involve copyright infringement, it is primarily an ethical issue [16].

Y.H. Zhang, *Against Plagiarism*, Qualitative and Quantitative Analysis of Scientific and Scholarly Communication,
DOI 10.1007/978-3-319-24160-9_1

The Oxford English Dictionary defines plagiarism as 'The practice of taking someone else's work or ideas and passing them off as one's own'; interestingly, it also gives the origin of the word as 'Early 17th century: from Latin *plagiarius*, 'kidnapper' (from *plagium*, 'a kidnapping', from Greek *plagion*)…' [17]. As mentioned in the Foreword, plagiarism is nothing new; the Roman poet Martial complained about it in the first century CE.

The Publication Ethics Committee of WAME (World Association Medicine Editors) provides a more detailed definition: 'Plagiarism is the use of others' published and unpublished ideas or words (or other intellectual property) without attribution or permission, and presenting them as new and original rather than derived from an existing source. The intent and effect of plagiarism is to mislead the reader as to the contributions of the plagiarizer. This applies whether the ideas or words are taken from abstracts, research grant applications, Institutional Review Board applications, or unpublished or published manuscripts in any publication format (print or electronic).' Pagiarism is scientific misconduct and should be addressed as such [18].

In 2010, the Committee on Publication Ethics (COPE) organized an annual seminar on the theme of plagiarism, at which Harold Sox, former Editor of the *Annals of Internal Medicine*, said 'Plagiarism is an act of fraud involving both stealing information and lying about where information came from' [19].

The United States Office of Research Integrity Policy on Plagiarism states that 'plagiarism includes both the theft or misappropriation of intellectual property and the substantial unattributed textual copying of another's work. It does not include authorship or credit disputes' [20].

In their book, *The Handbook of Journal Publishing* [6], Sally Morris et al. describe plagiarism as follows: 'Plagiarism is the use of someone else's work without attribution—in other words, passing it off as one's own. Text, figures, tables, and even ideas can be plagiarized. When a whole entity (e.g., an entire article, a figure, a table, or a dataset) is republished without attribution or permission, there may be a copyright violation as well as ethical misconduct.'

Many universities provide clear guidance for students and faculty on their websites about academic standards, including codes of conduct for authors (which naturally incorporate advice on how to avoid plagiarism). The guidance on Oxford University's website states: 'Plagiarism is presenting someone else's work or ideas as your own, with or without their consent, by incorporating it into your work without full acknowledgement. All published and unpublished material, whether in manuscript, printed or electronic form, is covered under this definition. Plagiarism may be intentional or reckless, or unintentional. Under the regulations for examinations, intentional or reckless plagiarism is a disciplinary offence' [21]. Harvard University's site states clearly 'In academic writing, it is considered plagiarism to draw any idea or any language from someone else without adequately crediting that source in your paper. It doesn't matter whether the source is a published author, another student, a Web site without clear authorship, a Web site that sells academic papers, or any other person: Taking credit for anyone else's work is stealing, and it is unacceptable in all academic situations, whether you do it intentionally or by accident' [22].

Given all of the authoritative definition of plagiarism, we can't help but want to ask who would be in the groups prone to plagiarism?

Who Commits Plagiarism, and Why?

In recent years, there have been a number of high-profile cases, as a result of which the perpetrators have variously lost their jobs, degrees and academic reputations [23–26]. At the same time, in some parts of the world plagiarism still goes largely unpunished; Vasily Vlassov, of the Russian Society for Evidence-Based Medicine, has pointed out that some prominent Russians, including Vladimir Putin himself, have been suspected of plagiarism [19]. Our research, republished in the following chapters, suggests that plagiarism is rife in both developed and developing, and Anglophone and non-Anglophone, countries. Although two recent articles [27, 28] both identified that that researchers from industrialized/developed counties 'tend to plagiarize less often than researchers elsewhere', and that 'there have been high profile cases of misconduct from low-and middle-income countries', the impact of cases from developed countries [23–30] is perhaps particularly damaging; they should be setting a global example based on the fact that copyright has a well-documented 300-year history based on law in Western countries [31] whereas it has been established much more recently in developing countries (e.g. in China the international copyright law has been in effect only since 1991) [32].

Plagiarism is clearly on the increase. As Bernard Rous, director of publications at the Association of Computing Machinery, commented, 'Not so many years ago, we got one or two alleged cases a year. Now we are getting one or two a month' [33].

So why do people (often people from the educated elite of society) plagiarize? Why would someone risk their reputation by committing such serious misconduct? There are a number of possible reasons for plagiarism, including the following:

- Increased pressure to publish
- Ease of copying and pasting online work
- Difficulties in writing in English (or another language)
- Misplaced respect for others' work
- Lack of suitable training
- Lack of awareness of the rules for acknowledgement of others' work

Increased Pressure to Publish

Intellectuals have always been that element of society that both creates spiritual wealth and promotes scientific progress. In the course of history, science and technology have always been among the key productive forces advancing the progress of human society. So it is understandable, perhaps, that scholars who want to gain prestige in this inviting field can find it difficult to resist the temptation to cheat in order to reach their goals.

Ease of Copying and Pasting Online Work

The Internet makes it temptingly easy to cut and paste sections—even large sections—of pre-existing publications. This habit can begin at school, or at university. Miguel Roig, a psychology professor at St John's University, observes that in the university campus, 'while student cheating, particularly plagiarism, should be old news… it continues to challenge the integrity of undergraduate and graduate education' [34]. In the same article, he shares the results of a large-scale study by Donald McCabe of Rutgers University which clearly illustrates the scope of the problem. In McCabe's study, 'of more than 70,000 undergraduates and 10,000 graduate students surveyed across 83 college campuses in the United States and Canada, 62 % of undergraduate and 59 % of graduate-student respondents admitted to have engaged in cut-and-paste plagiarism from online sources' [35].

Difficulties in Writing in English (or Another Language)

Globalization is an inevitable trend in today's academic journals. English has become established as the *lingua franca* of the sciences and, indeed, many of the arts and humanities, and thus most of the world's prestigious mainstream journals—not just in Anglophone countries, but also elsewhere [36]—are published in English. Researchers therefore prefer to present their work in English in order to reach the maximum audience. If English is not their native language, this may be difficult for them; one can understand (though not condone) the temptation to find phrases, sentences and more which express, in clear and accurate English, the ideas that a non-native speaker may be struggling to express.

Misplaced Respect for Others' Work

In some cultures, copying is considered a mark of respect for the original author. Not only even as Vlassov in his 'Cultural differences in plagiarism' [19] of COPE's 2010 annual seminar reported '...while in Asia, using the words of others can perhaps be linked to their tradition of respect for authority,' but also as Thatcher noted, 'Chinese consider copying, not as a disreputable or dishonest activity, but as the traditional way for the individual to learn and gain mastery of a field. In China, students are taught that to become better writers they need to memorize and imitate the language and style of heralded past masters; such appreciation for the achievements of a long cultural tradition is supposed to help them become more creative, too. And it is not just students who are imbued with such ideas. Many of their professors (...) practice plagiarism in the belief that emulation of experts in their fields is the way to advance in their careers' [32].

Lack of Suitable Training/Lack of Awareness of the Rules for Acknowledgement of Others' Work

While a growing number of universities now provide online guidance for their students (and faculty) on academic publication ethics, including plagiarism (see Chap. 8 for more details), this is still not universal. Because plagiarism, particularly cutting and pasting, is so easy to do in the Internet age, many young people (and even their elders) find it hard to see why it is wrong. Education about the fact that plagiarism involves both stealing from the original author, and lying about the provenance of their content, needs to start from early childhood.

The aim of increasing protection of scientists' original work and respect for authors' copyrights must therefore be a long-term one; promoting correct academic behavior and putting a stop to academic misconduct cannot be completed overnight. Developing a culture, throughout the academic and scientific world, which recognizes the key importance of integrity and honesty in one's own work, and the respect for the work of others, is a long-term undertaking.

What Is the Role of Journal Editors?

Ensuring the integrity and quality of their publication is one of the most basic responsibilities of every academic journal editor.

It is not pleasant for journal editors to feel that every submitted item must be suspected of potential plagiarism. However, any published article which contains plagiarized content brings discredit on the journal, and on its editor, as well as on the offending author. It is therefore the journal editor's duty, however painful, to carry out thorough checks, in order to ensure that no plagiarized material slips through.

The first step is for journal editors to be aware of the problem, to follow accepted codes of practice (see Chap. 10) and to ensure that plagiarism is always appropriately acted upon. Journal editors must be able to identify plagiarism when it occurs, to determine the nature and severity of the plagiarism, and then to take appropriate action.

Detecting Potential Plagiarism

Journal editors should use appropriate tools to detect potential plagiarism even though we know the tools can't tell the difference between genuine plagiarism and permissible duplication. Since the advent of search engines such as Google, Baidu, etc. it has been possible to conduct a search using a section of suspected content. However, it is not possible to enter the complete text of an article in the search box; in addition, the duplicated text may not be freely accessible, and of course there is no reporting support. Various plagiarism detection tools, such as Turnitin [37], have been available since the early years of the 21st century. However, the first tool specifically designed for the academic publishing market, CrossCheck (powered by iThenticate) [38] was launched in June 2008. In the same year, CrossCheck won the ALPSP Award for Publishing Innovation, which recognised its effective role in helping publishers to prevent academic fraud and to protect authors' copyrights by reducing the incidence of plagiarism. Since its advent, CrossCheck has been incorporated into the standard editorial process of many journals.

To February 2015, CrossCheck's underlying database included more than 126,260 journal titles and over 42 million individual articles (or other content items), from over 642 publishers; in excess of 190,000 additional manuscripts are uploaded every month (most of them in English) [39]. However, there are many other academic publications in different languages and scripts, all of which face the same problem of how to detect and prevent plagiarism. In China, for example, Chinese-language journals have been able to police plagiarism since 2009, using the AMLC (Academic Misconduct Literature Check) software developed by China Academic Journals Electronic Publishing House and Tongfang Knowledge Network Technology Co. Ltd. [40]. In addition, the CrossCheck system is now able to examine texts in a number of other languages including French, German, Korean, and Japanese (although it is only able to detect duplication in the same language) [32].

Types of Plagiarism

Oxford University's website (mentioned above) defines 8 forms of plagiarism and other authorial misconduct (we have indicated with an asterisk those which involve forms of plagiarism):

- 'Verbatim (word for word) quotation without clear acknowledgement*
- Cutting and pasting*
- Paraphrasing*
- Collusion
- Inaccurate citation
- Failure to acknowledge assistance
- Use of material written by professional agencies or other persons
- Auto-plagiarism*' [7].

In 2015, iThenticate published the interesting results of a survey on researcher behavior [41]; their respondents identified no fewer than ten separate types of plagiarism and other attribution issues (again, we have marked with an asterisk those which involve forms of plagiarism):

- Secondary source
- Invalid source
- Duplication*
- Paraphrasing*
- Repetitive research
- Replication*
- Misleading attribution
- Unethical collaboration
- Verbatim*
- Complete*

In our own journals (as discussed in detail in Chap. 10) we have identified nine types of plagiarism:

- Self- (or team) plagiarism without identification and acknowledgement
- Cutting and pasting of others' work without identification and acknowledgement
- Replication of methods sections (in Biomedical journals) without clear statement of the source
- Republication of conference papers with little added value
- Review papers which largely replicate previously published content
- Plagiarism of images/tables/formulae/data without both acknowledgement and copyright permission
- Plagiarism of ideas
- Wholesale plagiarism of previously published text
- Republication in translation without acknowledgment, permission and full citation.

Our experience as editors, and our own research [42], suggests that journal editors tend to pay the greatest attention to certain types of plagiarism: cut-and-paste, duplication of conference proceedings, self-plagiarism, team plagiarism, and review articles containing excessive amounts of quotation from the cited original papers. It is therefore important that, having studied the CrossCheck similarity reports and compared the submitted article with those with which it has a high similarity, the editor should decide what type of plagiarism (if any) he or she is dealing with, in order that the response may be appropriate.

Taking Appropriate Action

We have outlined in detail in Chap. 10 the guidelines and workflows that are available to journal editors from the most reputable international sources. We also provide there, for editors' consideration, an example in the form of the policies and workflows that we have adopted for our own journals [43].

Academic journal editors around the world have a huge responsibility—they are in the position of being able to detect, and therefore to stop plagiarism. While CrossCheck and other similar tools have become invaluable for journal editors, however, all they can actually do is highlight strings of similar text. Plagiarism cannot be accurately identified solely by finding these strings—closer examination is always needed. And the conclusion reached, and action taken, may differ between disciplines and in different countries.

In the chapters that follow, therefore, we examine how journal editors worldwide actually use CrossCheck at the moment, and their attitudes to different types of plagiarism. It is clear from our studies that editors do not yet always take as rigorous a position as one might wish, but this highlights the need to continue educating both researchers and editors at all levels about this important issue.

References

1. Feng, Youlan. 2006. *A short history of Chinese philosophy*. Beijing: Beijing University Press.
2. Cicero Quotes (607 quotes). http://izquotes.com/quote/282137.
3. Bosman, P. 2012. Corruption and integrity in ancient Greece and Rome. In Classical Association of South Africa, ed. P. Bosman. Pretoria.
4. Worth a 1000 words. 2010. *Ethical editing*. 2(4). http://publicationethics.org/files/u661/EthicalEditing_Winter2010_final_2.pdf.
5. Comment, Nature. 2012. How to stop plagiarism. *Nature* 481(21–23): 2012.
6. Morris, S, E. Barnas, D. LaFrenier, and M. Reich. 2013. *The handbook of journal publishing*, 1st ed., 360. Published in the United States of America by Cambridge University Press.
7. Meddings, K. 2010. Credit where credit's due: Plagiarism screening in scholarly publishing. *Learned Publishing* 23: 5–8.
8. Hames, I. 2007. *Peer review and manuscript management in scientific journals: Guidelines for good practice*, 1st ed, 177–181. New York: Wiley-Blackwell.

9. Roig, M. 2010. Plagiarism and self-plagiarism: What every author should know, Special issue: Responsible writing in science. *Biochemia Medica* 20(3): 295–300.
10. Zhang, YH, and XY, Jia. 2012. A survey on the use of crosscheck for detecting plagiarism in journal articles. *Learned Publishing* 25(4): 292–307.
11. Zhang YH, XY, Jia, HF, Lin, and XF, Tan. 2013. Editorial: Be careful! Avoiding duplication: A case study. *Journal of Zhejiang University-SCIENCE B (Biomedicine & Biotechnology)* 14(4): 355–358.
12. Jia XY, XF, Tan, and YH, Zhang. 2014. Replication of the methods section in biosciences papers: Is it plagiarism? *Scientometrics* 98: 337–45.
13. Zhang, Y.H., and X.Y. Jia. 2013. Republication of conference papers in journals? *Learned Publishing* 26(3): 189–196.
14. Zhang XX, ZL, Huo, and YH, Zhang. 2014. Detecting and (not) dealing with plagiarism in an engineering paper: Beyond. CrossCheck—a case study. *Science and Engineering Ethics* 20: 433–43.
15. ESF-ORI First World Conference on Research Integrity: Fostering Responsible Research. 2007. http://www.esf.org/index.php?id=4479.
16. Saunders, J. 2010. Plagiarism and the law. *Learned Publishing* 23(4). www.ingentaconnect.com/content/alpsp/lp/2010/.../art00002?.
17. Oxford English Dictionary Online. 2012. 3rd. ed. Oxford: Oxford University Press. Available at http://www.oxforddictionaries.com/definition/english/plagiarism.
18. World Association of Medical Editors, Recommendations on Publication Ethics Policies for Medical Journals. http://www.wame.org/about/recommendations-on-publication-ethics-policie#Plagiarism.
19. Seminar attendees ponder plagiarism. 2010. *Ethical editing* 2(2). http://publicationethics.org/files/u661/EthicalEditing_Summer2010_final.pdf.
20. ORI Policy on Plagiarism. https://ori.hhs.gov/ori-policy-plagiarism.
21. Oxford Students. *Academic matters: Plagiarism.* http://www.ox.ac.uk/students/academic/guidance/skills/plagiarism.
22. Harvard Guide to Using Sources. http://usingsources.fas.harvard.edu/icb/icb.do.
23. Pidd, H. 2011. German defense minister resigns in PhD plagiarism row. *The guardian*, March 1, 2011. http://www.theguardian.com/world/2011/mar/01/german-defence-minister-resigns-plagiarism.
24. German education minister quits over PhD plagiarism. *The guardian*, February 9, 2013. http://www.theguardian.com/world/2013/feb/09/german-education-minister-quits-phd-plagiarism.
25. Hungarian President Pal Schmitt stripped of doctorate. *BBC news*, March 30, 2012. http://www.bbc.com/news/world-europe-17558166.
26. A bold move against plagiarism. *China daily*, August 27, 2014. http://usa.chinadaily.com.cn/opinion/2014-08/27/content_18495472.htm.
27. Bohannon, J. 2014. Study of massive preprint archive hints at the geography of plagiarism. *Science*. Available from http://news.sciencemag.org/scientific-community/2014/12/study-massive-preprint-archive-hints-geography-plagiarism (accessed December 11, 2014).
28. Ana, J., T. Koehlmoos, R. Smith, and L.L. Yan. 2013. Research misconduct in low- and middle-income countries. *PLoS Med* 10(3): e1001315. doi:10.1371/journal.pmed.1001315.
29. Harvard plagiarism archive: Proof that laurence tribe used ghostwriters on the second edition of his treatise, August 29, 2010. http://authorskeptics.blogspot.com.
30. Will Harvard law school condemn Elizabeth Warren for pow wow chow plagiarism? http://www.breitbart.com/big-government/2012/05/20/will-harvard-discipline-warren/.
31. Williams, A. 2007. Copyright: past, present and future. *Learned Publishing* 20: 113–119. doi:10.1087/174148507X185072.
32. Thatcher, Sanford G. 2008. China's copyright dilemma. *Learned Publishing* 21(4): 2008.
33. Butler, D. 2010. Journals step up plagiarism policing cut-and-paste culture tackled by CrossCheck software. *Nature* 466: 167. http://www.nature.com/news/2010/100705/full/466167a.html.

34. Roig, M. 2010. *Plagiarism: An ounce of prevention… ASQ higher education brief*, June 2010. http://asq.org/edu/2010/06/continuous-improvement/plagiarism-an-ounce-of-prevention-.pdf.
35. McCabe, D.L. 2005. Cheating among college and university students: A North American perspective. *International Journal for Educational Integrity* 1(1). www.ojs.unisa.edu.au/index.php/ijei/article/viewfile/14/9.
36. Habibzadeh F, and H. Marcovitch. 2011. Plagiarism: the emperor's new clothes. *European Science Editing* 37(3). http://www.ease.org.uk/sites/default/files/eseaug11essayhabibzadeh.pdf.
37. Turnitin. http://en.wikipedia.org/wiki/Turnitin.
38. COPE. Plagiarism screening: An update on CrossCheck. *Ethical Editing* 2(2). http://publicationethics.org/files/u661/EthicalEditing_Summer2010_final.pdf.
39. Pentz E. 2015. CrossCheck: 2015 CrossRef workshop in Shanghai, April 29, 2015. http://www.crossref.org/crosscheck/index.html.
40. Chinese software detects academic plagiarists. *XinhuaNet*, March 17, 2009. http://news.xinhuanet.com/english/2009-03/17/content_11026942.htm.
41. iThenticate. 2015. 10 types of plagiarism in research, April 2015. http://www.ithenticate.com/resources/papers/decoding-plagiarism?utm_campaign=Non-branded%3A+Plagiarism&utm_source.
42. Zhang YH (Helen). 2011. A global survey on detecting plagiarism in journals using CrossCheck/Ithenticate, September 2011 in 2011 CrossRef Annual Meeting (PDF 745 kb).
43. Zhang YH (Helen), HF, Lin, and Q, Ye. 2015. Anti-plagiarism policy of *JZUS-A/B & FITEE*. *J Zhejiang Univ-Sci A (Appl Phys & Eng)* 16(6): 507–12. http://www.zju.edu.cn/jzus/opentxt.php?doi=10.1631/jzus.A1500151.

Chapter 2
Differences Between Anglophone and Non-anglophone Journals

As mentioned in the Preface, in 2011 the author and her colleagues received a grant from the Committee on Publication Ethics (COPE) in support of a research study intended to develop evidence-based guidance for journal editors on how to deal with different kinds of plagiarism detected through the use of CrossCheck [1]. The first part of the research was a global survey of journal editors and publishers, the results of which were published in an article entitled 'A survey on the use of Cross Check in detecting plagiarism in academic journals', which appeared in *Learned Publishing* in 2012 [2]. The article is reproduced in its entirety (with the publisher's permission) below.

Introduction

Although there are many dictionary definitions of plagiarism [3–5], Hames [6] provides the stricture that 'duplication of text or results from other articles or books is clearly unacceptable, either from the work of other authors or from an author's own work (known as auto- or self-plagiarism)'. Plagiarism or unreasonable levels of copying in journal articles is a growing problem: 'Not so many years ago, we got one or two alleged cases a year. Now we are getting one or two a month' [7]. 'How to stop plagiarism' [8] is an important topic in academic publication. The first step is for journal editors to use some tools to detect real plagiarism to protect authority and originality. In recent years, CrossCheck has become 'an established part of the editorial process for many journals' [9–13]. And as of February 2012, 291 publishers together accounting for more than 10,000 journals have become members of CrossCheck [14].

However, in our experience, CrossCheck is just a useful tool to help the editor find strings of similar text. Most instances of true plagiarism cannot be identified solely by these strings. This led us to question: (1) How do journal editors

Y.H. Zhang, *Against Plagiarism*, Qualitative and Quantitative Analysis of Scientific and Scholarly Communication,
DOI 10.1007/978-3-319-24160-9_2

worldwide use CrossCheck and handle the similarity reports? (2) What are journal editors' attitudes and tolerance toward types of plagiarism in different disciplines and different countries? (3) What are mainstream views and differences to these problems between editors in native English-speaking countries and non-native English-speaking countries? With this in mind, we undertook a survey as part of a research project funded by the Committee on Publication Ethics [15] (questionnaire in Appendix 1).

Methodology

Survey Design

From May to June 2011, an online survey was carried out, using SurveyMonkey (http://www.surveymonkey.com). The questionnaires were sent in two separate groups:

1. Survey Version 1 (SV1, the full questionnaire), consisting of 22 questions, was sent to 3305 recipients mostly from Anglophone countries (1371 academic journal editors from CrossCheck members, 1263 academic journal editors from Nature Publishing Group, Oxford University Press, Cambridge University Press, and some scholarly societies, etc., and 671 recipients from the attendees of the 33rd Society for Scholarly Publishing Annual Meeting, 2011).
2. Survey Version 2 (SV2, part of the full questionnaire), consisting of 10 questions (marked with * before numbers of questions in Appendix 1) chosen from SV1 because most of SV2 recipients without CrossCheck membership would not have been able to respond to all of the SV1 questions, was sent to 607 non-native English speaking editors of academic journals covered by Web of Science, from Japan, South Korea, India, Singapore, China and Brazil etc. (Fig. 2.1a).

We mostly sent the invitations to editors-in-chief and scientific editors although we had not asked the specific role of these journal editors in the questionnaire. (In other words, our questionnaire went to academics, subject experts who are directly involved in their respective journals and are likely to responsible for selection of content. In Asia (e.g. China) the managing editor performs many of the roles of the scientific editor in the West. The scientific editor/editor-in-chief is more of a figurehead so the questionnaire could have reached some managing editors.)

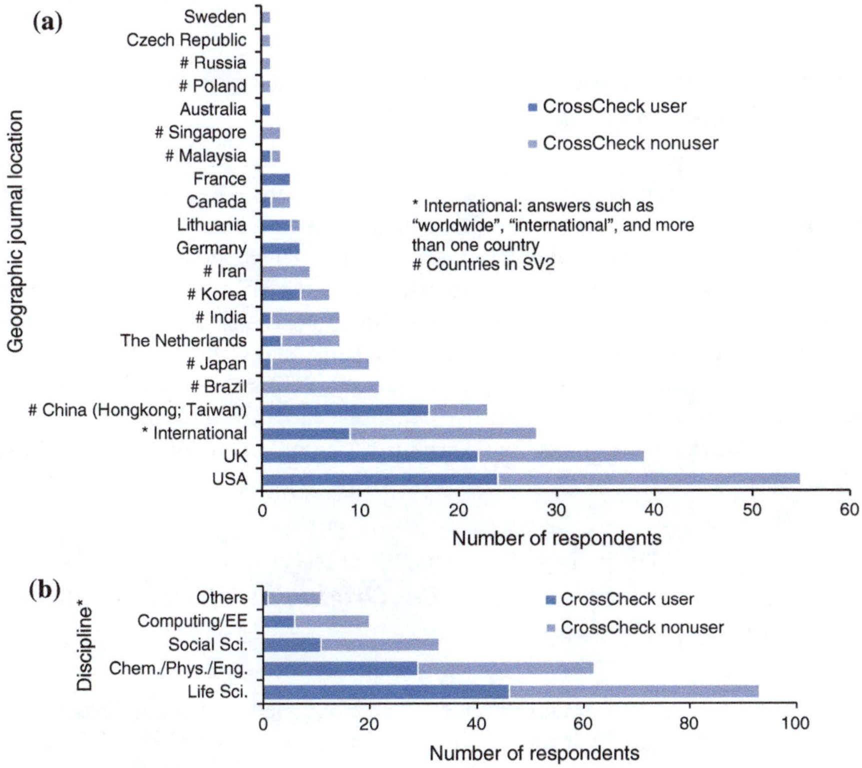

Fig. 2.1 All respondents analyzed by geographic journal location (**a**) and by discipline (**b**), showing CrossCheck users and nonusers (Q1 and Q2 cross-analyzed with Q3, SV1 and SV2, $n = 219$). *Life Sci. includes Bio-Sciences, Medicine and Agriculture, etc.; Chem./Phys./Eng. includes Mechanical/Civil/Environmental/Industrial/Control Engineering, Aerospace, Architecture, Mathematics and Statistics, etc.; Social Sci. includes Anthropology, Economics, Education, Geography, History, Law, Linguistics, Political Science, Public Administration, Psychology and Sociology, etc.; Computing/EE includes Computer Science, Electronics, Electrical Engineering (EE), Automation and Artificial Intelligence, etc.

Data Analysis

For multiple choice questions, the percentage of each option was calculated and compared, and the results were cross-analyzed, as appropriate, by native language of the respondents, by discipline and by geographic journal location. The answers to Q6 and Q7 were analyzed using discriminant analysis (classification analysis) to classify original data, and finally to calculate their mean and standard deviation, respectively. And for Q9, Q16, Q21 they were calculated as the mean or median.

Results

Respondents by CrossCheck Users, Journal Locations, and Disciplines (SV1 and SV2, Q1–Q3)

The number of respondents to SV1 was 161, mainly from Western countries. Most of their journals are from some top publishers, such as Nature Publishing Group, IEEE, Elsevier, Springer, Wiley-Blackwell, Oxford University Press, Cambridge University Press, and some scholarly societies from the USA, which include some leading journals, such as *New England Journal of Medicine and The Lancet* (see full list in Appendix 2). And 95 % of respondents show that their journals are published in English. There were 58 respondents to SV2; most of their journals are also among the leaders in their respective countries, and 93 % of these are published in English. Overall, the response rate was 5.6 %, from 21 countries (Table 2.1 and Fig. 2.1a).

Use of CrossCheck in Checking the Originality of Submitted Articles (SV1, Q4)

In SV1 and SV2, 42 % of total respondents have experience of using CrossCheck: 51 % of respondents to SV1, and 19 % of respondents to SV2 (Table 2.1). In the field of Life Sciences, 49 % respondents are CrossCheck users (Fig. 2.1b). In addition, a few respondents indicated that they use a variety of other methods to detect possible plagiarism, such as eTBLAST, Medknow's plagiarism checking tool, Free Online Plagiarism Checker.

As Table 2.2 indicates, 32 % of those SV1 respondents who use CrossCheck screen all submissions, while 34 % screen only suspect papers, and 15 % screen only accepted papers. The preference for screening only suspect papers is more

Table 2.1 Respondents to the survey (SV1 and SV2)

	Number of invitations	Number of respondents		Response rate (%)	Percentage of CrossCheck users among respondents (%)
		CrossCheck users	CrossCheck non-users		
Survey version 1	3305	82	79	4.9	51
Survey version 2	607	11	47	9.6	19
Total	3912	93	126	5.6	42

Table 2.2 Which papers are screened using CrossCheck (Q4, CrossCheck users in SV1, $n = 82$)

Disciplines	Percentage of options			
	All submissions (%)	Only suspect papers (%)	Only accepted papers (%)	Other (%)
Chem./Phys./Eng. ($n = 26$)	38	50	4	8
Life Sci. ($n = 39$)	31	28	18	23
Computing/EE ($n = 5$)	20	40	20	20
Social Sci. ($n = 11$)	27	18	27	27
Others ($n = 1$)	0	0	0	100
All ($n = 82$)	32	34	15	20

marked in Chemistry/Physics/Engineering etc. and Computer Sciences/Electrical Engineering, etc. than in Life Sciences and Social Sciences. Under the heading "Other", some editors stated that CrossCheck could be used at any time if a paper aroused suspicion; others said that they use it to check only non-research articles that are almost without figures, equations and tables, or long papers. Other editors indicated that the screening was done by the publisher before submissions reached them; one mentioned that "corresponding authors can choose to run their papers through CrossCheck (this is totally optional), and the journal pays the bill."

The respondents in SV1 and SV2 are also analyzed by different disciplines as shown in Fig. 2.1b which shows the replies from Life Sciences (Life Sci.) account for 43 %, Chemistry/Physics/Engineering, etc. (Chem./Phys./Eng.) for 28 %, Social Sciences (Social Sci.) for 15 %, Computer Science/Electrical Engineering, etc. (Computing/EE) for 9 %, and other disciplines for 5 %, respectively.

Use of CrossCheck Similarity Reports (Q5–Q7)

Figure 2.2 indicates how respondents use the CrossCheck similarity report. In SV1, 66 % of respondents supplement it with expert opinions; 20 % reject a paper out of hand if it is found to have an unacceptably high similarity (Table 2.3 indicates what similarity index respondents consider to be "unacceptably high"); 10 % forward the CrossCheck report to the reviewers for their advice if it gives rise to suspicions of possible plagiarism, while 4 % contact the authors to request an explanation.

Before giving the responses to Q6 and Q7, we need to define some aspects of the similarity report. The overall similarity index [16] (OSI) is the "percentage of similarity between a submission and information existing in the iThenticate databases selected as search targets. The single match similarity index [17] (SMSI) is the percentage of similarity from a single source between a submission and information existing in the iThenticate databases selected as search targets. The OSI

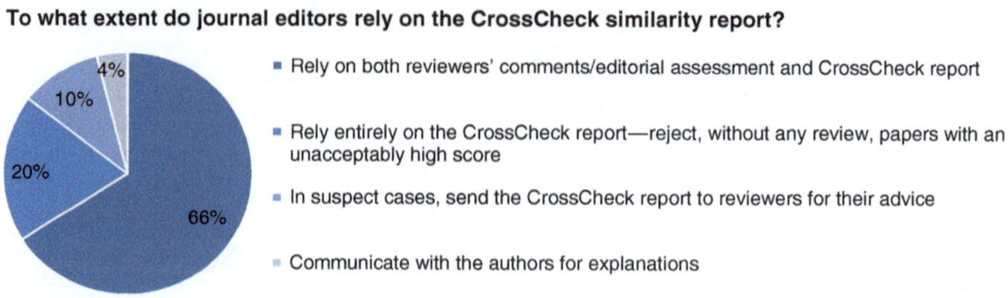

Fig. 2.2 How respondents use the CrossCheck similarity report (Q5, CrossCheck users in SV1, $n = 82$)

Table 2.3 How respondents view the significance of the similarity index (Q6 and Q7, CrossCheck users in SV1)

Seriousness (plagiarism/copying)	Suspected OSI ($n = 51$)		Suspected SMSI ($n = 46$)	
	Mean (%)	Standard deviation (%)	Mean (%)	Standard deviation (%)
Minor	8.99	4.23	8.99	4.23
Moderate	21.69	5.65	21.69	2.38
Serious	38.78	10.77	38.78	10.78
Trigger a reject	50.49	13.35	43.42	14.66
Trigger a rework	17.60	9.92	13.96	6.76

is one important indicator of a potentially plagiaristic paper; however, the degree of SMSI is also the other significant indicator. Table 2.3 shows the seriousness level of similarity that respondents feel suggests minor, moderate or serious plagiarism. In the case of the OSI, the mode was 8.99 ± 4.23 %, 21.69 ± 5.65 %, and 38.78 ± 10.77 %, respectively; the mode index which would trigger outright rejection was 50.49 ± 13.35 %, while that for requesting revision by the author was 17.60 ± 9.92 %. In the case of the SMSI, the mode was 8.99 ± 4.23 %, 21.69 ± 2.38 %, and 38.78 ± 10.78 %, respectively (the results for the SMSI were much higher than expected, which may indicate that respondents were not clear what the SMSI is).

Attitudes to Copying, Cutting and Pasting (Q8–Q14, Q19 and Q20)

Verbatim (or Near-Verbatim) Copying (Q8–Q10)

Q8 deals with verbatim (or near-verbatim) copying of an extract from another work. As shown in Fig. 2.3, about 60 % of respondents said that this can be acceptable

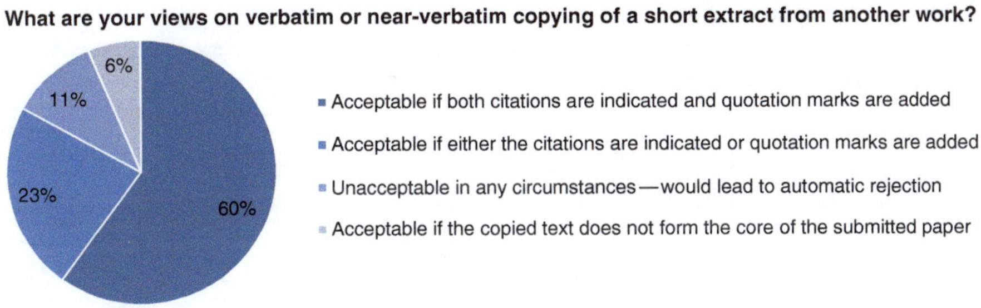

Fig. 2.3 Respondents' views on verbatim, or near-verbatim, copying of a short extract from another work (Q8, SV1, n = 160)

provided it is clear that it is a quotation (e.g. quotation marks, indentation) and a full citation is given for the original source; 23 % consider it acceptable with either quotation marks or a citation; 6 % find it acceptable if the copied text does not form the core of the submitted paper, while 11 % find it totally unacceptable and would always reject the paper.

Q9 deals with the length of word strings considered acceptable for verbatim copying, with or without citation. The data are cross-analyzed by discipline (Table 2.4). Respondents from Social Sciences show the lowest tolerance for word strings copied without citation. Respondents from the Life Sciences show considerably less latitude than other disciplines in the length of a quoted extract considered acceptable.

Q10 asks about the policy regarding authors who cut-and-paste materials from other sources and integrate this with their own text. From Fig. 2.4, it can be seen that 57 % of respondents indicated that this would be unacceptable in all cases and the paper would be rejected, but 23 % consider that cutting and pasting is acceptable if the paper is innovative, provided the author adds proper citations. Respondents from the Social Sciences are almost twice as likely to reject as those from the Life Sciences (Fig. 2.4b). Chinese respondents are particularly disinclined to reject in these circumstances (Fig. 2.4c).

Table 2.4 Length of extract (number of words) considered acceptable for verbatim copying with and without citation (Q9, SV1, n = 138)

Disciplines	Number of words (median)[a]	
	Without citation	With citation
Chem./Phys./Eng. (n = 42)	10	50
Life Sci. (n = 54)	8	30
Computing/EE (n = 14)	10	50
Social Sci. (n = 28)	1	50
All (n = 138)	7	50

[a]Data for the number of words are shown as a median as the responses varied widely

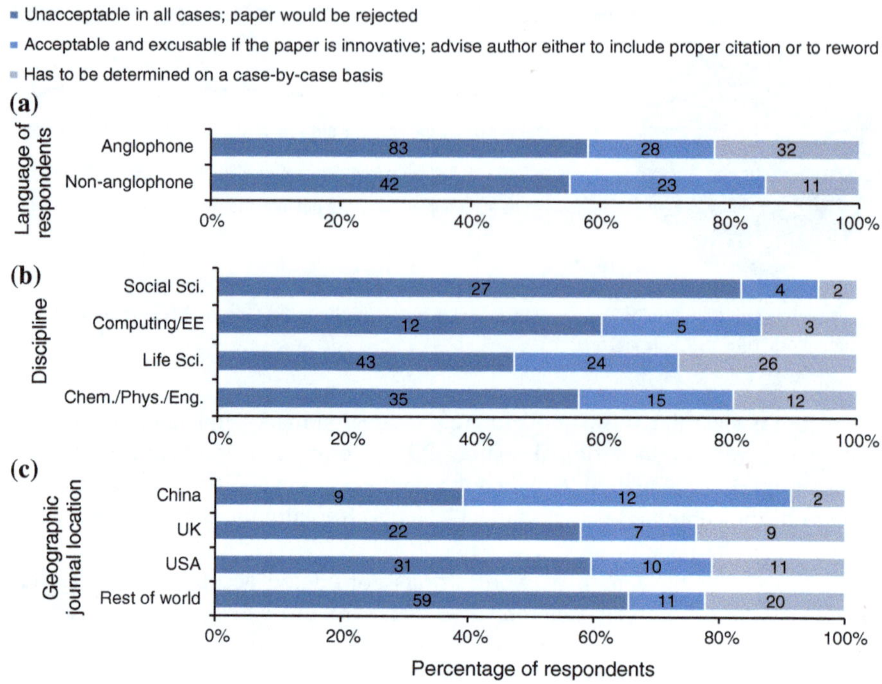

Fig. 2.4 Policy of respondents regarding authors who cut-and-paste materials by language of respondents (**a**), by discipline (**b**), and by geographic journal location (**c**) (Q10, SV1 and SV2, $n = 219$). In this figure and all subsequent bar chart the number in bars shows the actual numbers of respondents

Attitude and Tolerance to Copying in Different Sections of a Paper (Q11–Q14)

Q11 and Q13 on editors' attitudes to the copied materials occurred in different parts of the articles. In Fig. 2.5, the majority of respondents indicated that if 1/4–1/3 of the content in the Abstract, Introduction or Discussion is copied without citation, the paper is likely to be rejected. Respondents from the Life Sciences and Chemistry/Physics/Engineering, etc. are less likely to reject than the other respondents; they prefer to ask the author to include a proper citation, or to rewrite the content in his/her own words.

When cut-and-paste occurred in the Materials & Methods section of a paper (Q13), respondents generally indicated that this was unacceptable unless rewritten using the author's own words or with proper citation (Fig. 2.6). Comments in response to this question suggested that decisions would be based on journal requirements (e.g., "some journals ask for a detailed description of the method"), article genre, type of text (e.g., "some technical points are very difficult to re-word

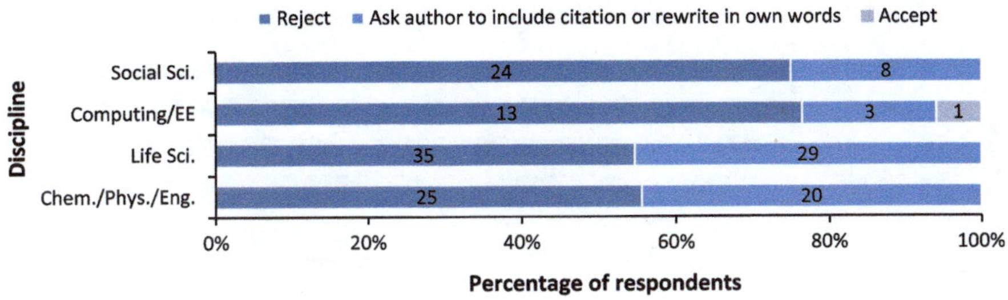

Fig. 2.5 Attitude to copied content in abstract, introduction or discussion, by discipline (Q11, SV1, total response $n = 161$ which includes data of other disciplines ($n = 3$) not shown in this chart)

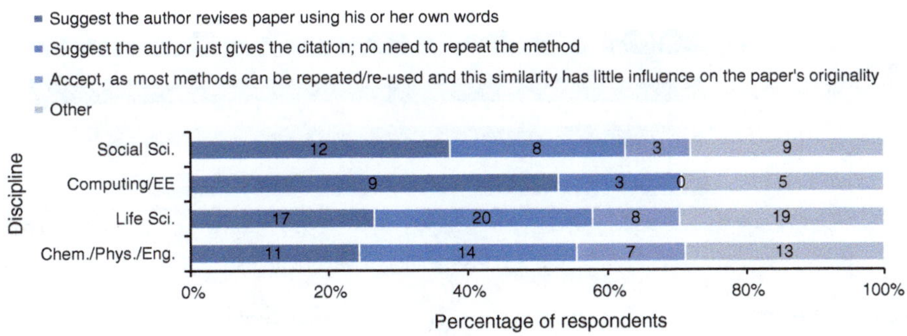

Fig. 2.6 Attitude to copied content in materials and methods section without citation, by discipline (Q13, SV1, total response $n = 161$ which includes data of other disciplines ($n = 3$) not shown in this chart)

while retaining their meaning. It is also difficult to tell the difference between whether the person is copying from person *X*, or if both are copying from another source, like a text book"), length of the text (e.g., "short string of purely technical detail about the methodology is OK"), or that advice would be sought from reviewers or the editor-in-chief.

Q12 and Q14 ask for editors' tolerance of copied content in the sections of Abstract/Introduction/Discussion and the section of Materials and Methods, respectively. In the case of Abstract, Introduction and Discussion, more than 90 % of respondents suggest that even with citation the acceptable percentage of copied content is very low, between 1 and 20 % (Fig. 2.7). In the case of the Materials and Methods section, about 70 % of respondents indicated that copied content should be 20 % or below. However, nearly 20 % (13/64) from Life Sciences would tolerate the copied contents of 21–40 % (Fig. 2.8).

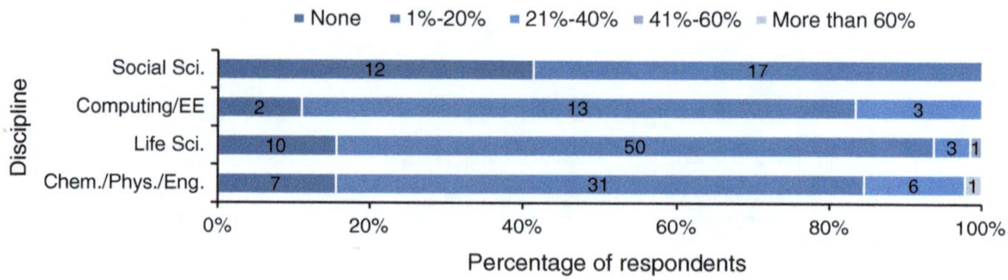

Fig. 2.7 Percentage of copied content in abstract, introduction or discussion sections considered acceptable with citation, by discipline (Q12, SV1, total response n = 161 which includes data of other disciplines (n = 5) not shown in this chart)

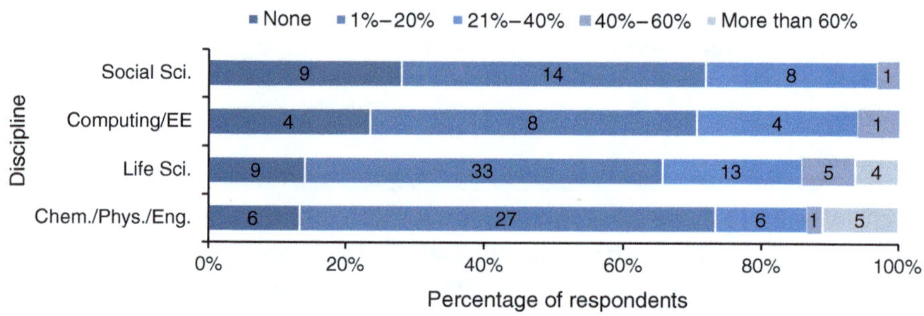

Fig. 2.8 Percentage of copied content in materials and methods section considered acceptable with citation, by discipline (Q14, SV1, total response n = 161 which includes data of other disciplines (n = 3) not shown in this chart)

Review Articles Consisting Predominantly of Copied Text 'Cutting-and-Pasting' (Q19 and Q20)

Summaries of the papers discussed, often using the original authors' own words, are unsurprisingly more common in review articles; Q19 and Q20 address the respondents' views about the acceptability of cutting-and-pasting in review articles. From the respondents in different disciplines and languages, we can see an average 79 % of them responding to Q19 suggested that even review articles would be rejected in their current form, or accepted only after rewriting in the review author's own words, if the summaries consisted wholly or mainly of the original authors' words. A total of 25 % of Anglophone respondents would reject such papers out of hand; for non-Anglophone respondents this percentage was 13 % (Fig. 2.9a). There were no marked disciplinary differences (Fig. 2.9b).

78 % of respondents to Q20 felt that review papers with an OSI > ~50 % would not be acceptable, and more than 60 % of respondents said that the acceptable OSI in a review article would be <35 %. Again, disciplinary differences were not marked (Fig. 2.10). 16 % choose "others" showing no metrics, which depends on what is similar and why.

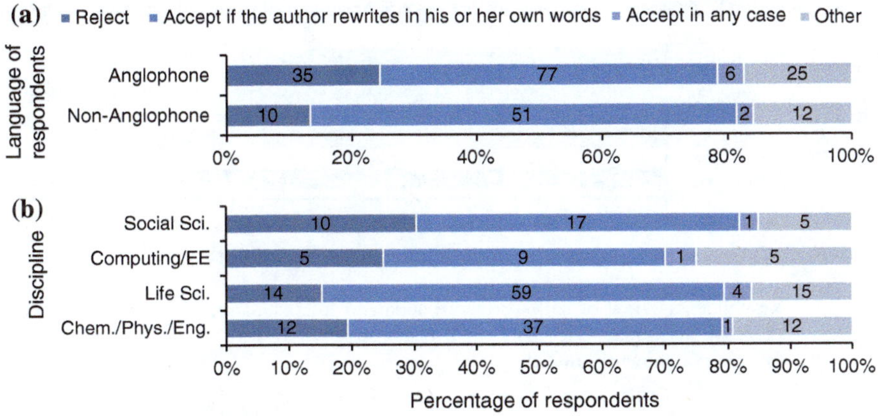

Fig. 2.9 Respondents' course of action when the author of a review article has summarized previously published papers wholly or mainly in the original authors' own words, by language of respondents (**a**) and by discipline (**b**) (Q19, SV1 and SV2, n = 218)

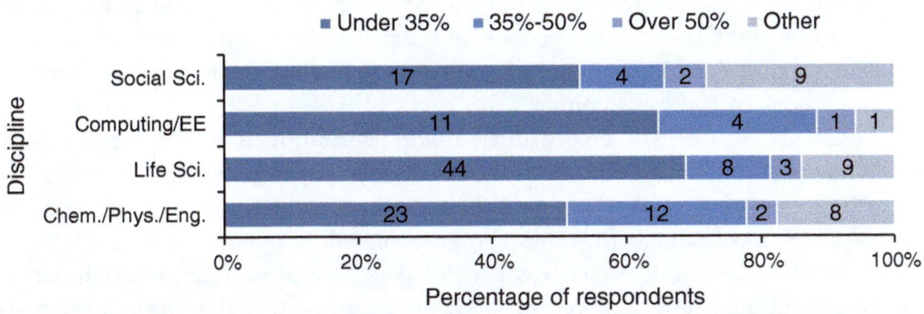

Fig. 2.10 Percentage figure for the overall similarity index considered acceptable in a review article, by discipline (Q20, SV1, total response n = 161 which includes data of other disciplines (n = 3) not shown in this chart)

Self-/Team Plagiarism and Duplicate Publication (Q15–Q18)

Self-plagiarism and Team Plagiarism

Is it ethical or reasonable that 'author or co-authors reuse their own previously written work or data in a 'new' written product without letting the reader know that this material has appeared elsewhere' [17]? Q15, Q17 and Q18 deal with self-plagiarism and team plagiarism, which means copying from each other a great deal within the same research program or group. While the definition of self-plagiarism is difficult, and although self-plagiarism does not involve the theft of someone else's work, it is still likely to contravene journal policy, and often also violates the original publisher's copyright [18]. Plagiarism by an individual of

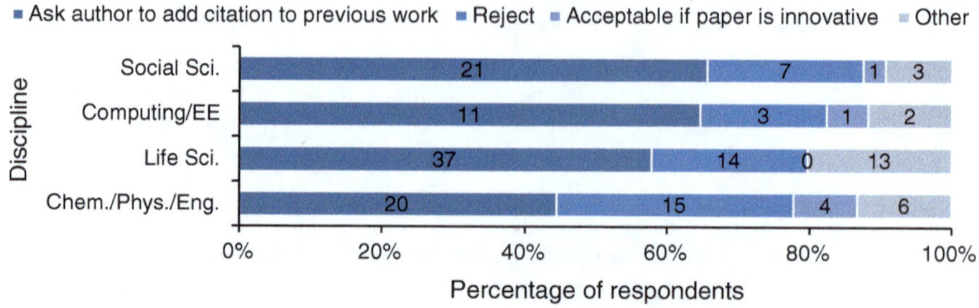

Fig. 2.11 Respondents' view of self- or team plagiarism in Results and Conclusions sections without citation, by discipline (Q15, SV1, total response n = 161 which includes data of other disciplines (n = 3) not shown in this chart)

his/her previous work (self-plagiarism) and plagiarism of the work of the team in which the author is a member are very similar; in both cases the author is copying his/her own or his/her group's previously published tables, figures and text with no or few changes, without making clear what has been copied and without citing the previous publication. The responses to Q15, which refers to self- or team plagiarism in the Results and Conclusions sections of a paper, indicate that the most common response is to ask the author to provide a citation to his/her previous work; there are few differences between disciplines (Fig. 2.11).

Q17 asked "How do you deal with an article whose title, aims and methodology are identical or highly similar to those of another paper published by the same research group (team), and where only the specific examples and materials, etc. are different?" 91 % of editors from both Anglophone and non-Anglophone countries responded that they would either reject such a paper out of hand, or accept it only if revised to highlight new findings or innovations, and citing the group's previous publication(s) (Fig. 2.12a). Far fewer respondents in Computer Science/Electrical Engineering were inclined to reject out of hand than those in other disciplines (Fig. 2.12b). Some answers mentioned other factors, such as the referee's opinion, the author's explanation, and the amount of significant additional content.

Q18 asks how the respondent would react if the author(s) claim that the papers are a series of studies with the same background, which will inevitably lead to similarity in the text. 44 % of respondents said they would accept but only with citation, while 24 % said they would reject with or without citation. Irrespective of language and discipline (Fig. 2.13), there is a strong consensus (total of 68 %) either to reject, or to accept only with proper citation. A number of respondents commented that the author must rephrase or revise with proper citation, minimizing repetition, and eliminating all cutting and pasting to avoid self-plagiarism. One respondent mentioned that the reviewer's or editor-in-chief's comments on the innovation or originality of the paper should also be taken into account; others stressed that if the similarity is in a core area of the article (such as in the results section) with no substantially new overall contribution, the article should be rejected out of hand.

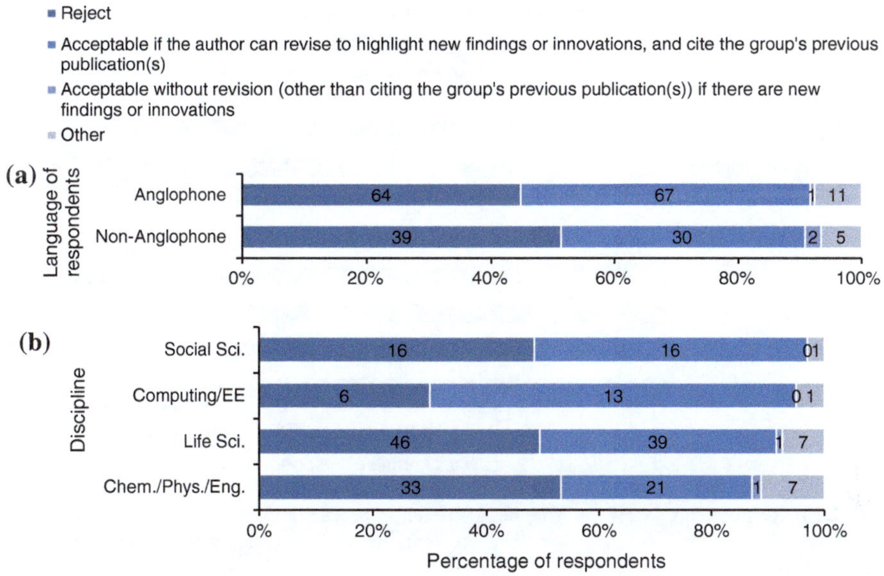

Fig. 2.12 Respondents' view of suspected team plagiarism, by language of respondents (**a**) and by discipline (**b**) (SV1 and SV2, Q17, $n = 219$)

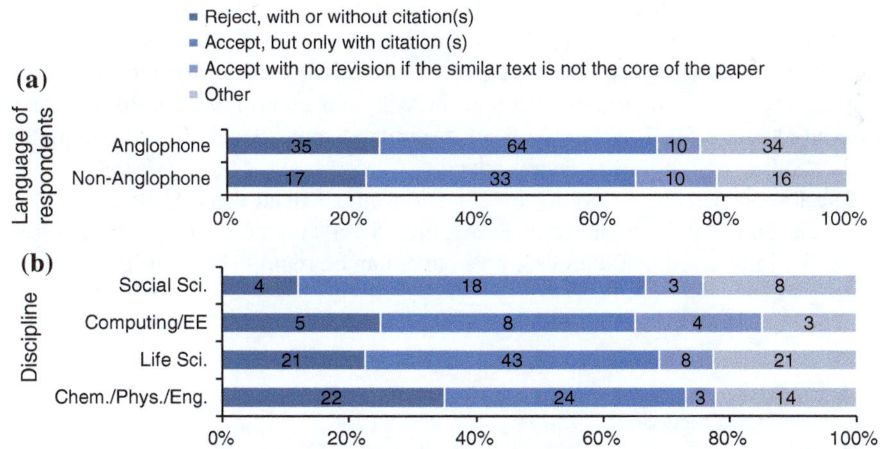

Fig. 2.13 Respondents' course of action in cases of significant self-plagiarism, where the author (s) claim that the papers are a series of studies, by language of respondents (**a**) and by discipline (**b**) (Q18, SV1 and SV2, $n = 219$)

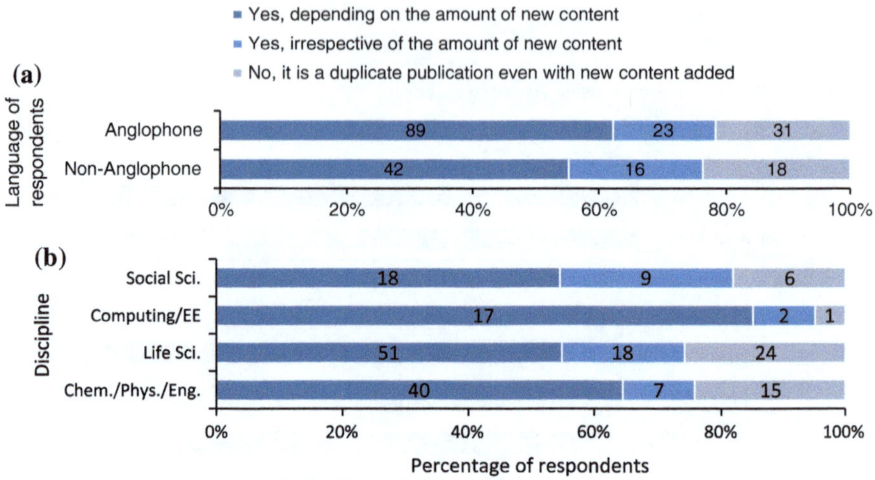

Fig. 2.14 Respondents' view on whether conference proceedings papers can legitimately be republished in a journal, by language of respondents (**a**) and by disciplines (**b**) (Q16, SV1 and SV2, n = 219)

Republication of Papers from Conference Proceedings (Q16, SV1 and SV2)

Q16 addresses the question "Should papers previously published in conference proceedings legitimately be republished in journals?" 60 % of respondents think such papers can properly be republished provided they include new content (Fig. 2.14a, b). And they indicated that there should be no less than 46 % of new material (Table 2.5). However, 22 % of respondents considered this to be duplicate publication, even with new content added.

The lowest percentage opting for rejection out of hand came from the field of Computer Science/Electrical Engineering, etc. (5 %), and more than 80 % of editors in this field indicated that a proceedings paper can be republished if it includes new content.

Table 2.5 Amount of new content considered necessary by respondents in order to justify republication of papers from conference proceedings (Q16, SV1 and SV2, n = 131)

Disciplines	n	Percentage of new content (mean) (%)
Chem./Phys./Eng.	40	50
Life Sci.	51	45
Computing/EE	17	35
Social Sci.	18	45
Others	5	43
All	131	46

Table 2.6 Percentage of papers submitted to respondents' own journals that are rejected specifically on account of plagiarism (Q21, SV1 and SV2, $n = 153$)

Respondents by language	n	Percentage rejected specifically on account of plagiarism (mean) (%)
Anglophone	102	3.5
Non-anglophone	51	11.0
All	153	5.9

Percentage of Papers Rejected Because of Plagiarism (Q21)

Q21 asks respondents to give the approximate percentage of papers rejected because of plagiarism. From the answers to Anglophone ($n = 102$) and non-Anglophone ($n = 51$) editors, we find the rejection rates depending on plagiarism based on each journal's experience are about 3.5 and 11.0 %, respectively (Table 2.6).

Willingness to Refer to Use of CrossCheck in 'Instructions to Authors' (SV1, Q22)

Q22 asks whether respondents would be willing to refer to their use of CrossCheck in the 'Instructions to Authors'. As shown in Table 2.7, 45 % of respondents (SV1, $n = 148$) indicated that they already did so; a further 28 % would consider doing so. However, 19 % thought that it would be intimidating and counter-productive, and 8 % would not (or not yet) consider such a statement.

Respondents' approach to suspicion of plagiarism varies relatively little between journals from developed and developing countries, and between journals from Anglophone and non-Anglophone countries, even though the percentage of articles rejected on the grounds of plagiarism by journals from non-Anglophone countries is almost double that of journals from Anglophone countries.

Table 2.7 Respondents' willingness to state in their "Instructions for Authors" that they use CrossCheck to scan all papers submitted (Q22, SV1, $n = 148$)

Option	Percentage
Already state that we use CrossCheck to scan for plagiarism	45
Would consider stating that we use CrossCheck	28
Do not (or not yet) state that we use CrossCheck	8
Would not state that we use CrossCheck: seems intimidating and counter-productive	19

Discussion

What Is and Is not Acceptable

'There are tools to detect non-originality in articles, but instilling ethical norms remains essential' [19]. The Merriam-Webster Dictionary defines plagiarize as 'to commit literary theft: present as new and original an idea or product derived from an existing source' [6]. A number of guides to research publication [7, 20–22] make it clear that taking text (even a phrase or a sentence) from someone else's work without acknowledgement is theft of their intellectual property—plagiarism, whether it is deliberate or not. The survey data suggest a strong consensus among all respondents about the criteria for determining plagiarism; this was more marked than the local variations (whether based on language, discipline or geographical journal location).

Respondents to the survey generally agree that verbatim (or near-verbatim) copying must include a citation and direct quotation marks, and that no more than 7 words should be copied without citation and quotation marks, and the strictest response is in social sciences, only one word (Table 2.4). Respondents in Chemistry/Physics/Engineering, etc. and in Computer Science/Electrical Engineering and Social Sciences are of the opinion that even if a citation and quotation marks are given, no more than 50 words should be copied; respondents in the Life Sciences felt that citations of up to 30 words were acceptable.

In a review article, the author aims to digest a wide range of previous papers and synthesize her/his findings to form a coherent argument about a topic or a focused description of a field. 78 % of respondents felt that review papers with an OSI > ~50 % would not be acceptable, and 60 % felt that those with an OSI > 35 % would not be acceptable (Fig. 2.10).

Even when repeating 'Common Knowledge' [23, 24] in some sections of a paper, such as classical methods in biomedical procedures, respondents felt that authors should wherever possible use their own words. Only one respondent felt that 1/3–1/4 cut-and-pasted content could be acceptable in the Abstract, Introduction or Discussion sections (Fig. 2.5), but 11 % felt that this was acceptable in the Materials and methods sections (Fig. 2.6).

There is a great temptation to plagiarize one's own work or that of one's team, since the number of publications is often used as an indication of a researcher's scientific merit [25]. However, team plagiarism has long been criticized by journal editors and publishers [26–28]. In this study, in the opinion of all groups, whether by discipline, language or geographic location, over 90 % of editors were highly consistent in their disapproval of team plagiarism.

As to whether self-plagiarism can be defined as a type of plagiarism, there are a lot of discussions in the iThenticate White Paper [18]. And in our survey, it is encouraging that the majority of editors would either reject a substantially self-plagiarizing article (24 %), or accept it but only with the addition of proper citations (44 %). And to this question "is it possible to steal from oneself?" Hexam pointed out "the essence of self-plagiarism is the author's attempts to deceive the reader" [29].

Re-publication of papers which have previously appeared in conference proceedings is a difficult issue; there are circumstances in which re-publication of papers (with the permission of the copyright owner) has been considered acceptable in the past. However, there is now a huge amount of information available via the Internet. Having similar or duplicate content can cause confusion and waste publishing resources. For example, in academic journal publishing, as far as we know most journals are published online with digital object identifiers (DOIs). A re-published paper with the same content could have two DOIs that will result in misleading information and waste the reader's time. 60 % of respondents to the survey felt that proceedings papers could properly be re-published in journals provided they contained at least 46 % of new content (Fig. 2.14; Table 2.5).

Differences by Discipline and Language

Disciplinary Differences in Rejection Decisions

For five typical plagiarism problems such as cut-and-paste (Q10), duplication of conference proceedings (Q16), self-plagiarism (Q17), team plagiarism (Q18), review papers of high similarity (Q19), based on the rejection decisions' percentage of respondents from these subjects, we can find differences between different disciplines (Fig. 2.15).

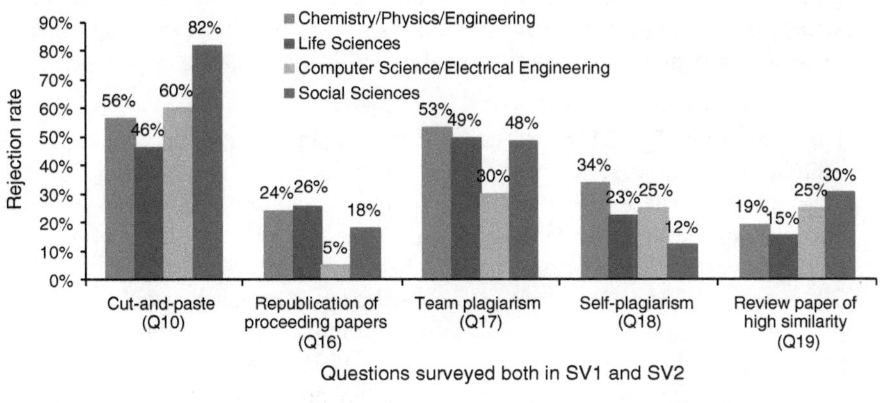

Fig. 2.15 Differences in those choosing "rejection" in response to 5 typical questions in different disciplines

We find that the Social Sciences show the lowest tolerance, with 82 and 30 % rejection rates, respectively to cut-paste (Q10) and much copied material in review papers (Q19), but to self-plagiarism (Q18) they show a high tolerance, with a rejection rate of only 12 %. What is the reason? This may be related to the characteristics of social science articles, with greater expression of a personal viewpoint in the text. Perhaps because the writing process in the social sciences is itself a creative one, then social science editors express the lowest tolerance here. Further exploration of self-plagiarism may be needed. As the iThenticate White Paper on the ethics of self-plagiarism [16] states: 'writers maybe unaware of the ethics and laws involved in reusing or repurposing their own texts', although our journal editors do show they recognize the problem by using the plagiarism detecting tool.

The other noticeable phenomenon is that Computer Science/Electrical Engineering shows the lowest rejection rate in republication of proceedings papers (Q16), and team plagiarism (Q17), with 5 and 30 %, respectively. Because this subject mainly depends on the updating of new technologies and team cooperation, there are more conference proceedings publications.

The attitude toward the five questions in Chemistry/Physics/Engineering and Life Sciences disciplines seems to be very similar.

Language Differences

(a) Majority consensus
Figure 2.16a illustrates the extent of consensus between respondents from English speaking countries (mainly in SV1, n = 143) and non-native English speaking countries (mainly in SV2, n = 76), to a number of key questions. Especially for team plagiarism, both groups show over 91 % "reject" rate or 'acceptable if can revise to highlight new findings or innovations, and cite the group's previous publication(s)'. And the percentages of the other four mainstream questions also are also over 55 %, which proves global editors have expressed a strong mainstream view about ethical standards.

(b) Difference in minority opinions
However from Fig. 2.16b, we know there are small variations that show the attitude of non-Anglophones is a little less rigorous than that of the Anglophones. These differences may be due to cultural and language differences arising from the wide range of social perspectives and stages of national development [30]. Copyright law has been well-documented for more than 300 years in Western countries [31, 32], whereas it has been established much more recently in developing countries (for example, in China the international copyright law has been in effect only since 1991). Thus, some authors from developing countries may be unaware that they are committing plagiarism and infringing copyright law [32]. It may take some time before authors from

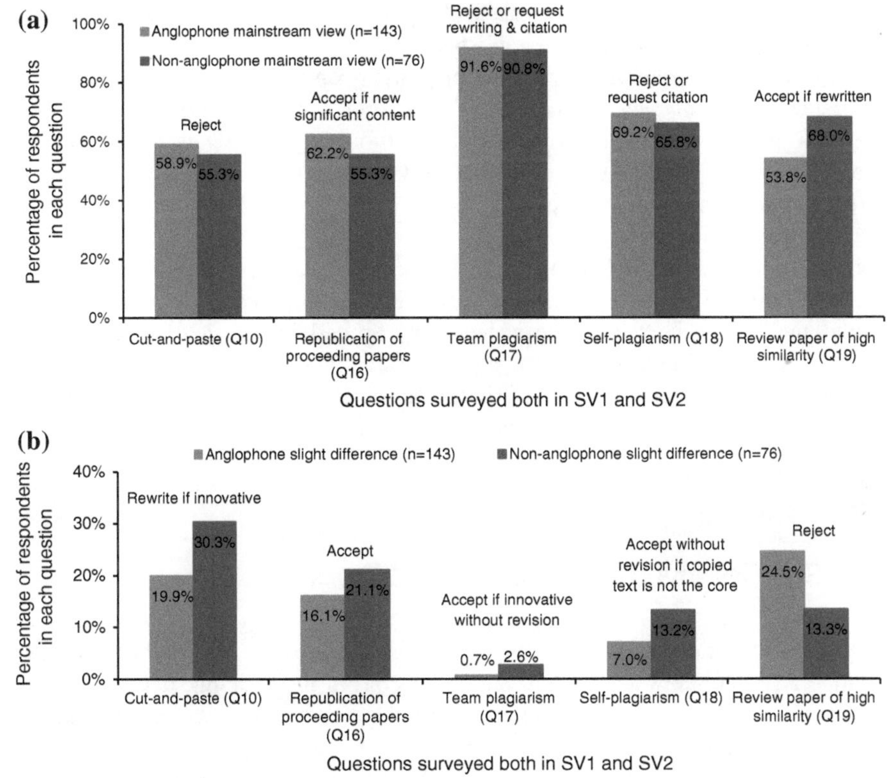

Fig. 2.16 Mainstream view (**a**) and small differences (**b**) to five problems between Anglophone (n = 143) respondents and non-Anglophone (n = 76) respondents

developing countries catch up with Western countries, and before the incidences of plagiarism can effectively be reduced in those countries.

Conclusions

The main results of this survey can be summarized as follows: (1) The plagiarism detection tool and similarity report are very useful and effective, and can assist editors to screen documents suspected of plagiarism. (2) Global editors have expressed a strong mainstream view of ethical standards even though there are slight variations between different disciplines and countries, as well as between non-Anglophone editors and Anglophone editors. (3) Given those variations perhaps a global principle and practical approaches to prevent plagiarism and duplicate publication should be established.

Any attempt to reduce or prevent plagiarism and duplicate publication will require that not only academic journal editors or scholarly publishers, but also scientists themselves, whether acting as authors or as reviewers, accept the responsibility to raise their own standards, and indeed to establish criteria so that the next generation can clearly understand the difference between ethical and unethical publishing behavior [27].

In addition, our survey indicates a relatively high awareness of plagiarism issues, and high ethical standards in dealing with them, among academic journal editors around the world. However, the survey was limited in that it only approached relatively few leading journals and reported on a small sub-set of global journals. In some ways undertaking a survey itself can be seen as a form of advocacy for higher ethical standards; it may be worthwhile undertaking a more comprehensive survey.

Practical Lessons for Authors and Editors

Plagiarism detection tools, such as CrossCheck, and the similarity reports they produce are extremely useful and can assist editors in screening documents suspected of plagiarism; a large number of journals already use them. However, these tools have limitations: they cannot search non-textual content (such as figures, tables, graphs etc.), and neither can they identify plagiarism of ideas. At the same time, they may identify text which is similar for perfectly acceptable reasons. Editors have to exercise their professional judgement (as well as relying on that of their reviewers) in order to determine whether any type of plagiarism has, in fact, occurred.

Our survey results indicate a clear overall global consensus on editorial standards on plagiarism. However, there were small variations between disciplines and between countries, as well as between Anglophone and non-Anglophone journals. This does not mean that lower standards can or should be applied in certain areas. More work is needed to raise all journals' standards, worldwide, to those of the top international journals.

Appendices

The survey questionnaire and the breakdown of respondents were made available as an online supplement to this paper when it was published in Learned Publishing in 2012; they remain available at http://www.zju.edu.cn/jzus/download/editorpapers/Authorsversion.pdf.

Appendix 1: Survey Questionnaire (SV1 + SV2)

Title: Survey on Detecting Plagiarism in Journals using CrossCheck
#This survey is being carried out as part of a research project funded by the Committee on Publication Ethics (COPE).
(SV1 contains 22 questions, of which 10 (marked with *) were used in SV2)

***Q1 What is the subject of your journal(s)? Choose firstly a broad category, as below.**
Chemistry/Physics/Engineering (Mechanical/Civil/Environmental/Industrial/Control, Aerospace etc.), Architecture, Mathematics/Statistics
Life Sciences (including Bio-Sciences, Medicine, Agriculture)
Computer Science/Electronics/Electrical Engineering/Automation/Artificial Intelligence etc.
Social Sciences (Anthropology/Economics/Education/Geography/History/Law/Linguistics/Political Science/Public Administration/Psychology/Sociology)
Others

***Q2 Basic information about your journal(s)**
Country ______________
Language ______________

***Q3 Do you use CrossCheck?**
Yes
No

Q4 How do you use CrossCheck in checking the originality of submitted articles?
All submissions are CrossChecked
Only accepted papers are CrossChecked
Only suspect papers are CrossChecked
Other (please specify) ______________

Q5 As a journal editor, to what extent do you rely on the CrossCheck similarity report to judge whether submitted papers involve plagiarism?
Rely entirely on the CrossCheck report—reject, without any review, papers with an unacceptably high score
Rely entirely on reviewers' comments; do not consider CrossCheck report
Rely on both reviewers' comments and CrossCheck report
In suspect cases, send the CrossCheck report to reviewers for their advice
Other (please specify) ______________

***Q6 The overall similarity index percentage is one important indicator of a potentially plagiaristic paper.**
Please indicate at what percentage you decide the paper contains:
Minor plagiarism, minimum (%) ______________
Moderate plagiarism, minimum (%) ______________
Serious plagiarism, minimum (%) ______________
Triggers a reject, minimum (%) ______________
Triggers request to author to rework it, minimum (%) ______________
(for SV2, similar Q is "**In a journal paper, what percentage of copied content would you consider acceptable with citation?** ______________")

Q7 The degree of similarity for each single match is also significant. Please indicate for single matches at what percentage you decide the paper contains:
Minor plagiarism, minimum (%) ______________
Moderate plagiarism, minimum (%) ______________
Serious plagiarism, minimum (%) ______________
Triggers a reject, minimum (%) ______________
Triggers request to author to rework it, minimum (%) ______________

Q8 What are your views on verbatim or near-verbatim copying of a short extract from another work?
Acceptable if the copied text does not form the core of the submitted paper
Acceptable if both citations are indicated and quotation marks are added
Acceptable if either the citations are indicated or quotation marks are added
Unacceptable in any circumstances—would lead to automatic rejection

Q9 What length of extract (number of words) would you consider acceptable for verbatim copying in the following two cases?
Without citation (number of words) ______________
With citation (number of words) ______________

***Q10 What is your policy regarding authors who cut-and-paste materials from other sources and integrate it with their own text?**
Acceptable and excusable if the paper is innovative; advise author either to include proper citation or to rewrite in own words
Unacceptable in all cases; paper would be rejected
Other (please specify) ______________

Q11 In sections Abstract/Introduction/Discussion, if between 1/4 and 1/3 of the content is copied without citations, what would you do?
Reject
Ask author to include citation or rewrite in own words
Accept

Q12 In sections Abstract/Introduction/Discussion, what percentage of copied content would you consider acceptable with citation?
None
1–20 %
21–40 %
41–60 %
More than 60 %

Q13 In section Materials & Methods, if between 1/4 and 1/3 of the content is copied without citations, what would you do?
Accept, as most methods can be repeated/re-used and this similarity has little influence on the paper's originality
Suggest the author revises paper using his or her own words
Suggest the author just gives the citation; no need to repeat the method
Other (please specify) _______________

Q14 In section Materials & Methods, what percentage of copied content would you consider acceptable with citation?
None
1–20 %
21–40 %
41–60 %
More than 60 %

Q15 In section Results and Conclusions, what is your view of authors copying their own previously published tables or figures with no or small changes without citation?
Reject
Ask author to add citation to previous work
Acceptable if paper is innovative
Other (please specify) _______________

***Q16 Do you think papers previously published in conference proceedings can legitimately be republished in a journal with the addition of new content?**
No, it is a duplicate publication even with new content added
Yes, irrespective of the amount of new content
Yes, depending on the amount of new content. Please indicate what amount of new content as a minimum percentage _______________

***Q17 How do you deal with an article whose title, aims and methodologies are identical or highly similar to those of another paper published by the same research group, and where only the specific examples and materials, etc. are different?**
Reject

Acceptable if the author can revise to highlight new findings or innovations, and cite the group's previous publication(s)
Acceptable without revision (other than citing the group's previous publication(s)) if there are new findings or innovations
Other (please specify) ______________

***Q18 Authors sometimes reuse significant portions of their own work, either verbatim or near-verbatim (self-plagiarism); they may claim that the papers are a series of studies with the same background, which will inevitably lead to similarity in the text. What is your reaction in cases like this?**
Reject, with or without citation(s)
Accept, but only with citation(s)
Accept with no revision if the similar text is not the core of the paper
Other (please specify) ______________

***Q19 In writing a review paper, authors necessarily summarize (and cite) previously published papers. How do you handle cases where they have predominantly used the original authors' own words?**
Reject
Accept in any case
Accept if the author rewrites in his or her own words
Other (please specify) ______________

Q20 In a review paper, what percentage figure for the overall similarity index would you accept?
Under 35 %
35–50 %
Over 50 %
Other (please specify) ______________

***Q21 In your own journal(s) and hence subject area, approximately what percentage of papers you receive are rejected specifically on account of plagiarism? (We are not asking for the title of your journal(s) to avoid any possible malevolent use of this information)**

____

Q22 In your "Instructions for Authors", would you consider announcing that you use CrossCheck to scan all papers submitted, or would this seem intimidating and counterproductive?

____

____

Appendix 2: List of Respondents to SV1, by Publisher (n = 161)

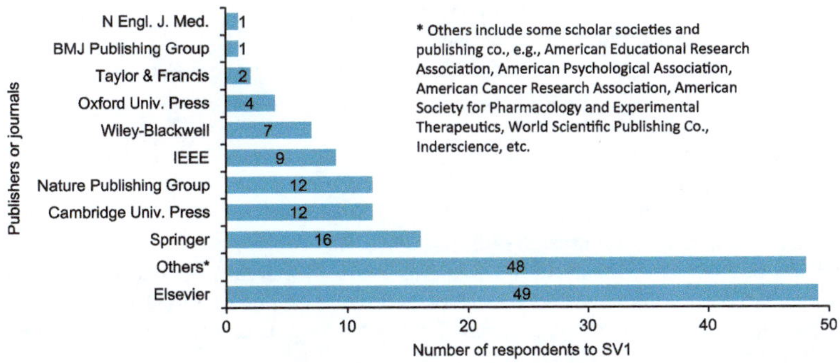

References

1. COPE Research Grant. 2010 December. CrossCheck guidance: an analysis of typical cases of plagiarism in different disciplines. http://publicationethics.org/resources/research.
2. Zhang, Y.H.(Helen), and X.Y. Jia. 2012. A survey on the use of CrossCheck in detecting plagiarism in academic journal articles. *Learned Publishing* 25(4). www.learned-publishing.org (with permission from Learned Publishing).
3. Plagiarism, n. 2012. *Oxford English dictionary online* (3rd Ed.). Available at http://www.oed.com.
4. Plagiarize, v. 2012. *Oxford English dictionary online* (3rd Ed.). Available at http://www.oed.com.
5. Plagiarize, n. 2012. *Merriam-Webster dictionary*. Available at http://www.merriam-webster.com/dictionary/plagiarize.
6. Hames, I. 2007. *Peer review and manuscript management in scientific journals: Guidelines for good practice* (1st Ed.), 177–81. Hoboken: Wiley-Blackwell.
7. Butler, D. 2010. Journals step up plagiarism policing cut-and-paste culture tackled by CrossCheck software. *Nature* 466: 167.
8. Nature Comments. 2012. How to stop plagiarism. *Nature* 481: 21–23.
9. Meddings, K. 2010. Credit where credit's due: plagiarism screening in scholarly publishing. *Learned Publishing* 23: 5–8.
10. Zhang, Y.H. 2010. CrossCheck: An effective tool for detecting plagiarism. *Learned Publishing* 23: 9–14.
11. Zhang, Y.H. 2010. Chinese journal finds 31 % of submissions plagiarized. *Nature* 467: 153.
12. Griffin, C. 2010. The journal of bone and joint surgery's CrossCheck experience. *Learned Publishing* 23: 132–135.
13. Kleinert, S. 2011. On behalf of the editors of all Lancet journals. Checking for plagiarism, duplicate publication, and text recycling. *The Lancet* 377: 281–82.
14. Available at http://www.crossref.org/crosscheck_members.html.

15. COPE Research Granted. 2011. CrossCheck guidance: An analysis of typical cases of plagiarism in different disciplines. Available at http://publicationethics.org/resources/research.
16. Available at http://www.crossref.org/crosscheck/CrossCheck_Manual.pdf.
17. Roig, M. 2006. Avoiding plagiarism, self-plagiarism, and other questionable writing practices: A guide to ethical writing. Available at http://facpub.stjohns.edu/~roigm/plagiarism/Self%20plagiarism.html.
18. iThenticate White Paper. 2011. The Ethics of Self-Plagiarism.
19. Nature Editorial. 2010. Plagiarism pinioned. *Nature* 466: 159–160.
20. Greetham, B. 2008. *How to write better essays* (2nd Ed.). London: Palgrave Macmillan.
21. McMillan, K., and J. Weyers., 2011. *How to write essays and assignments* (2nd Ed.). London: Pearson Education Limited.
22. Morris, S., E. Barnas, D. LaFeenier, and M. Reich. 2013. *Handbook of journal publishing*. Cambridge: Cambridge Univeristy Press.
23. Purdue University Online Writing Lab. 2010. Is it plagiarism yet? Available at http://owl.english.purdue.edu/owl/resource/589/02/ (accessed October 10, 2011).
24. Cal State San Marcos Library. 2008. How to avoid plagiarism? Available at http://library.csusm.edu/plagiarism/howtoavoid/how_avoid_common.htm (accessed October 10, 2011).
25. Franzoni, C., G. Scellato, and P. Stephan. 2011. Changing incentives to publish. *Science* 333: 702–703.
26. Benos, D.J., et al. 2005. Ethics and scientific publication. *Advances in Physiology Education* 29: 59–74.
27. Lin, H.F., et al. 2011. Guarantee academic originality: Duty of journal editors—Workflow and analysis with CrossCheck of *JZUS (A/B/C)*. *Chinese Journal of Scientific and Technical Periodicals* 22: 328–333. (in Chinese).
28. Radulescu, G. 1985. Duplicate publication is boring. *American Journal of Diseases of Children* 139: 119–120.
29. Hexham, I. 1999. *The plague of plagiarism*. Department of Religious Studies. The University of Calgary. Available at http://c.faculty.umkc.edu/cowande/plague.htm#self.
30. Suresh, S. 2011. Moving toward global science. *Science* 333: 802.
31. Williams, A. 2007. Copyright: past, present and future. *Learned Publishing* 20: 113–119.
32. Thatcher, S.G. 2008. China's copyright dilemma. *Learned Publishing* 21: 278–284.

Chapter 3
Publication in More Than One Language

For a variety of reasons, journal articles may be republished in translation (although this cannot be detected by computerized text-matching tools such as CrossCheck). It may be, for example, that the original language of publication is unlikely to reach one part of the author's intended readership. This raises the difficult question of whether or not this constitutes duplicate publication.

In 2013 we published an editorial on this topic [1] which is reproduced in full (with the publisher's permission) below.

Rationality and Transparency of Bilingual (Multilingual) Publication

The foreign language versions of the Bible, well-known as the world's best seller, had amounted to 2197 by the year 1997, which at least demonstrates that in the world there were more than 2000 written languages, based on data on www.baike.com. Similarly, among the published books in the world's commonly used languages, there must be translations of Einstein's "Theory of Relativity", showing that world culture and scientific communication depend on translation and publication in different languages, and thus provide the rationality for multilingual publication.

However, just like other activities of human beings, publishing activity involves ethics and morality problems. Since copyright laws were introduced, they have helped formulate the norms of publishing activity [2]. Recently, one article [3] from '*Science and Technology Review*' elaborated that in the process of Chinese and English bilingual submissions, authors should have credibility, i.e., non-infringement, and manuscripts should be easy to propagate. As a journal editor, combining the practice of publishing and the idea of publishing internationally, I would like to highlight the following points as references for journal editors handling and authors preparing for bilingual submissions, and a report on this topic, in Chinese, was published in *Science and Technology Review* [4].

Y.H. Zhang, *Against Plagiarism*, Qualitative and Quantitative Analysis of Scientific and Scholarly Communication,
DOI 10.1007/978-3-319-24160-9_3

1. *Transparency and permission*

Bilingual (multilingual) publication should include the following processes. After publishing one paper in a primary language, authors should get permission from the primary journal (some international journals have already provided an authorization statement in the text), and should state acknowledgement of a secondary publication or which parts are a secondary publication in the manuscript when submitting the manuscript in another language. Such action is regarded to be free of copyright suspicions, faithful to the primary version, and free of any academic misconduct. Then reviewers and journal editors will judge and decide whether the content of this manuscript is useful and necessary as a secondary publication or not.

2. *Achievement and evaluation*

After bilingual (multilingual) publishing, primary and secondary publications should be counted only once by relevant research or personnel departments in quantitative assessment of original research achievements by the identified author. This is because the primary publication shows the original research, while the secondary publication is just a translation of it. Of course translations also belong to creative works of language. For example, the translations of famous Chinese translators Mr. FU Lei and Mr. LIN Yu-tang have been of long-term cultural benefit to the readers. Therefore, the secondary publication may be a reference for language skills when relevant departments are assessing achievements.

3. *Indexing and referencing*

Indexing and referencing of primary and secondary publications are to be discussed, respectively. Indexing refers to databases. For the National Library of Medicine (NLM), when the primary and secondary publications both appear in the same journal, editors of journals should understand that MEDLINE/PubMed indexes the primary language version. For example, some Canadian journals publish manuscripts in both English and French simultaneously, the MEDLINE/PubMed database only indexes abstracts published in English, not in French, but both languages are indicated in the MEDLINE citation (such as [5, 6]). An index of the primal and secondary publications of other journals in different languages may vary, so duplicate abstracts might appear in the same database [7]. Reference of primary and secondary publications is usually based on the language preference of readers, so generally no special provisions exist.

Discussion of Duplicate Publications and Bilingual (Multilingual) Publications

As specified above, although bilingual (multilingual) publication is theoretically reasonable, it will undoubtedly have to be defined as a duplicate publication which is not to be commended if the publishing processes are not transparent.

Duplicate publications lead to double-counting in the assessment of research achievements, and moreover lead to fragmentation and misinterpretation of effective academic research, so it is contrary to ethical standards. What are the types of duplicate publication? What are the manifestations of such misconduct? In particular, under what circumstances is multilingual publication wrong? How to conduct this correctly? In one world, one principle of publishing, shall we learn from Table 3.1, named 'Guide to duplicate submission/publication and how to prevent it' from Elsevier [8] to discuss the rights and wrongs of duplicate publication and multilingual publication.

Analysis and correction of four types of duplicate publication behaviour clearly show that ethical standards of the international academic publishing industry are consistent. In particular, analysis of multilingual publications shows that without preconditions of transparency, multilingual publications might be suspected of infringement and misconduct, and then classified as duplicate publications.

Furthermore, I would like to communicate with my peers about our recent completed research findings from the Committee on Publication Ethics (COPE) Fund [9–14]. In particular, one investigation is that whether manuscripts which have been already published at computing/electronics conference proceedings, then submitted to journals, are duplicate submissions/publications? [11]. It was the consensus through the 1980s and 1990s in the field of computing and electronics that publishing of conference proceedings, prior to journal publication, contributed to rapid exchange of information, so resubmissions to journals were considered reasonable. However, with the rapid development of the Internet, concepts and techniques of the publishing industry have been updated, and the standards of ethics have been raised, so the norms of resubmissions have become suspect and there has been an attack on duplicate publication in recent years. Prof. Moshe Y. VARDI, the chief editor of the journal *Communications of the ACM* spoke out on this issue in the editors' forum: "we are driving on the wrong side of the publication road" [15]. Rome was not built in one day. Of course, changes in the norms for resubmissions need time, and consensus about stronger publishing ethics needs to be established and expanded.

Also, one typical case from COPE refers to a discussion of the rights and wrongs of duplicate publications [16]. This case involves four manuscripts in which three manuscripts were originally published in authors' native language and then the fourth was published in an English language journal. There was overlap in the authors who were involved in all four manuscripts and who were reported and suspected of duplicate publication. So the English language journal took the lead to investigate. Editors from different language journals collaborated with each other and contacted the authors to investigate and collect evidence, and finally decided to keep the third manuscript available and retract the other three papers, including the English one, due to duplication.

Furthermore, this case demonstrates that in the global village, the ethical standards of academia and the publishing industry are consistent. From another perspective, it also shows that nowadays journal editors should spend quite a lot of energy dealing with such problems, besides normal editing and publishing work. So attaching importance to academic ethics is the common responsibility of the scientific community and publishing industry.

Table 3.1 Guide to duplicate submission/publication and how to prevent it [8][a, b]

Action	What it is	Is it unethical?	What should you do?
Simultaneous submission	Submitting a paper to two or more journals at the same time	Yes Submission is not permitted as long as a manuscript is under review with another journal	1. Avoid submitting a paper to more than one publication at a time 2. Even if a submitted paper is currently under review and you do not know the status, wait to hear back from the publisher before approaching another journal, and then only if the first publisher will not be publishing the paper
Duplicate publication	When an author submits a paper or portions of his or her own paper that has been previously published to another journal, without disclosing prior submission(s)	Yes	1. Avoid submitting a previously published paper for consideration in another journal 2. Avoid submitting papers that describe essentially the same research to more than one journal 3. Always provide full disclosure about any previous submissions (including conferences, presentations and posting of results in registries) that might be regarded as duplicate publication 4. This should include disclosing previous publication of an abstract during the proceedings of conferences
Duplication by paraphrasing or "text-recycling"	When an author writes about his or her own research in two or more articles from different angles or on different aspects of the research without acknowledgment of the original paper	Yes Creating several publications from the same research, is considered manipulative. See our separate factsheet on plagiarism/text recycling	1. Put anything in quotes that is taken directly from a previously published paper, even if you are reusing something in your own words 2. Make sure to reference the source accordingly
Translations of a paper published in another language	Submitting a paper to journals in different languages without acknowledgment of the original paper	Yes Translated articles are acceptable when all necessary consents have been obtained from the previous publisher of the paper in any other language and from any other person who might own rights in the paper	1. If you want to submit your paper to a journal that is published in a different country or a different language, ask the publisher if this is permissible 2. At the time of submission, disclose any details of related papers in a different language, and any existing translation

[a]When in doubt, always consult with your professor, advisor, or someone in a position of authority who can guide you to the right course of action ([b]Reprinted with the permission of the Global Rights Department of Elsevier [1])

Practical Lessons for Authors and Editors

From the professional perspective of journal publishing, whether a bilingual (multilingual) publication is suspicious from an ethical point of view has been a topic of debate. Indeed this problem has been long puzzling authors who are preparing for submission. Although the worldwide publishing industry has had some discussions about it, so far in developing countries, for example, in China no normative text or guide is available for authors and journal editors as reference. In fact, it is not easy to sum up the rights and wrongs of duplicate publication and bilingual (multilingual) publication. From a professional perspective, a few highlights for discussion on the criteria for bilingual (multilingual) publication and substantial duplicate publication are proposed and could be a reference for authors and editors.

If an author re-publishes his/her own published work in translation without acknowledgment, permission and full citation, this is considered plagiarism.

Authors If an author submits for publication an article which has already been published in another language without acknowledging the fact, this would be considered self-plagiarism. The fact that the article has already been published in another language must be clearly acknowledged and the original publication fully cited; copyright permission must also be obtained from the original publisher.

Remedy—make clear that it is a translation, provide full citation to the original publication, and obtain written copyright permission from the original publisher.

Editors An article should only be accepted for republication in translation if the peer reviewers concur that this is in the best interests of the relevant scholarly community. Copyright permission must, however, be obtained from the original copyright owner, and a full citation must be given to the original publication.

References

1. Zhang, Y.H. 2013. Bilingual (multilingual) publications and duplicate publications: for or against? *Journal of Zhejiang University-SCIENCE A (Appl Phys & Eng)* 14(9): 687–690. doi:10.1631/jzus.A1300272. (Reproduction Permission from *JZUS*).
2. Williams, A. 2007. Copyright: past, present and future. *Learned Publishing* 20(2): 113–119. doi:10.1087/174148507X185072.
3. Zhang, Q. 2013. Reflection on moral issue of submitting the same writing for many publications in different languages. *Science & Technology Review* 31(4): 15–16. (in Chinese).

4. Zhang, Y.H. 2013. The rights and wrongs of bilingual publications and duplicate publications. *Science and Technology Review* 31(16): 17–18. (in Chinese).
5. Mercer, K. 2008. The relentless challenge in health care. *Healthcare Management Forum* 21 (2): 4–5. (English, French).
6. ICMJE. 2013. Uniform requirements for manuscripts submitted to Biomedical Journals: Overlapping Publications. Available from http://www.icmje.org/publishing_4overlap.html. Accessed on July 2013.
7. Errami, M., and H. Garmer. 2008. A tale of two citations. *Nature* 451(7177): 397–399. doi:10.1038/451397a.
8. Elsevier. 2012. Simultaneous submission/multiple, duplicate publication via ethics in research and publication Available from http://www.ethics.elsevier.com/pdf/ ETHICS_SSUB01a.pdf. Accessed on April 15, 2013.
9. COPE Research Granted. 2011. CrossCheck guidance: An analysis of typical cases of plagiarism in different disciplines. Available from http://publicationethics.org/resources/research. Accessed on January 26, 2011.
10. Zhang, Y.H., and X.Y. Jia. 2012. A survey on the use of CrossCheck for detecting plagiarism in journal articles. *Learned Publishing* 25(4): 292–307. doi:10.1087/20120408.
11. Zhang, Y.H., and X.Y. Jia. 2013. Republication of conference papers in journals? *Learned Publishing* 26(3): 189–196. doi:10.1087/20130307.
12. Zhang, Y.H., X.Y. Jia, H.F. Lin, and X.F. Tan. 2013. Editorial: Be careful! Avoiding duplication: A case study. *Journal of Zhejiang University-SCIENCE B (Biomedicine & Biotechnology)* 14(4): 355–358. doi:10.1631/jzus.B1300078.
13. Zhang, X.X., Z.L. Huo, and Y.H. Zhang. 2013. Detecting and (not) dealing with plagiarism in an engineering paper: beyond CrossCheck—A case study. *Science Engineering Ethics* 20(2): 433–443. doi:10.1007/s11948-013-9460-5.
14. Jia, XY., X.F. Tan, Y.H. Zhang. 2013. Replication of the methods section in biosciences papers: is it plagiarism? *Scientometrics* 98(1): 337–345. doi:10.1007/s11192-013-1033-5.
15. Vardi, M.Y. 2009. Conferences vs. journals in computing research. *Communications of the ACM* 52(5): 5. doi:10.1145/1506409.1506410.
16. COPE. 2011. Duplicate publication in possibly four papers. Case No. 11–21. Available from http://publicationethics.org/case/duplicate-publication-possibly-four-papers. Accessed on 20 April, 2013.

Chapter 4
How Not to Deal with It: A Case Study

Introduction

We describe here our own experience with plagiarism (which could not be detected by CrossCheck) in a particular article in the field of Engineering; our account was published under the title 'Detecting and (Not) Dealing with Plagiarism in an Engineering Paper: Beyond CrossCheck—A Case Study' [1] and is reproduced here in full, with the publisher's permission.

Figures, tables and formulae are key elements of research papers in fields such as engineering and the physical sciences. However, computational text-matching tools, such as CrossCheck, cannot detect potential plagiarism in these non-textual elements, although today, we have read the *Nature's* latest news, titled 'The image detective who roots out manuscript flaws' [2], and know EMBO (European Molecular Biology Organization)'s journals editors are checking all image-based scientific data in manuscripts by using the tool (Photoshop v CS2-CS3) provided by the US Office of Research Integrity (http://ori.hhs.gov/actions) [3] which may be the good news for journal publishers to prevent image aberrations and plagiarism in the future.

Consequently, when checking engineering or physical sciences papers, CrossCheck may return a low similarity index even when plagiarism has in fact taken place. The paper reproduced below discusses a case of plagiarism in engineering papers with a low similarity index from CrossCheck, together with some suggestions on how to tackle this type of problem.

Plagiarism Detection in Engineering and Physical Sciences Papers

CrossCheck is a powerful tool that enables editors to compare the text of a paper against a large database of published academic literature, as well as against material freely available on the internet (www.crossref.org/crosscheck.html) [4–7]. However,

Y.H. Zhang, *Against Plagiarism*, Qualitative and Quantitative Analysis
of Scientific and Scholarly Communication,
DOI 10.1007/978-3-319-24160-9_4

almost all plagiarism detectors, such as CitePlag, CrossCheck and PlagScan [8], can only decide whether a given body of scanned text exceeds a threshold of similarity when compared to another body of text [9]; human judgment is still needed to make the final decision on whether plagiarism has actually taken place. Furthermore, materials copied from other papers may have been reworded, or paraphrased, making any plagiarism difficult to detect by text-matching. Specially, in engineering and physical sciences papers, there is the added complication that non-textual elements—figures, tables, formulae and, more recently, videos and audio files—are often the basic elements that directly communicate the authors' ideas, concepts, methods (especially algorithms), data, results and inferences of a study. Such non-textual material may represent a significant portion of the content of a paper, even main intellectual component [10]. Thus, any plagiarism can only be found by examining and comparing these elements, as well as the text.

As Morris et al. [11] indicated that copyright provides the legal underpinning for much of what journal publishing does, Brinkman [12] and Wager [5] also stressed that the invention of various detecting tools is aimed at help journal editors protect copyright and safeguard academic integrity. Recently, due to the research grant of the Committee on Publication Ethics (COPE) with the aim of developing evidence-based guidance for journal editors on how to deal with different kinds of plagiarism detected through the use of CorssCheck [13], we have conducted several research surveys and published a few papers on what journal editors actually do in different subjects) [14–19].

As a computational text-matching tool [20], CrossCheck is unable effectively to find matches in figures, equations, tables, videos or audio files. The following is a typical case prompted us to think about how to confirm, deal with and avoid plagiarism in these areas.

Case Study

Background

It is not easy to use CrossCheck to identify matching formulae, tables, and figures, so a low similarity index report on engineering papers may not tell the whole truth. In this paper, we present an interesting case study on this matter. All persons involved have been anonymised.

The case began when Author A, who had identified and reported an alleged case of plagiarism involving one of his papers, wanted to obtain more data to support his claim. He discovered from correspondence published in *Nature* [13] that the present authors used CrossCheck to detect plagiarism. He therefore wrote asking them to help him check a paper written by Author B (with a co-author). Author A had previously published a paper (Paper A) on structural stability in Journal A. He noticed that a paper (Paper B) on a similar topic published a few years later by

Author B in Journal B took large sections of material from his earlier paper without adequately acknowledging the source.

Before contacting us, Author A had already performed a detailed analysis of the similarity between the two papers to support his case, putting similar sections of text and similar equations side-by-side in a table for ease of comparison. The table clearly showed that the piece of analysis carried out in Paper B was very similar to that published in his earlier paper, albeit with some slight modifications to take into account the interactions between neighboring structures. Although Paper B did cite paper A, Author A felt that it gave inadequate acknowledgement to his paper as the source of much of the material, especially the core formulae; the impression given from reading Paper B was that Author B did most of the analysis himself. As an independent check, the present authors used CrossCheck to compare Paper B against our database. If a paper has an overall similarity index (OSI:)[1] $>\sim 40\ \%$ or a single match similarity index (SMSI)[2] $>\sim 10\ \%$, we usually reject it out of hand at the initial submission [19]. The CrossCheck report gave an OSI of 32 %; the highest SMSI was about 5 %. There was no long paragraph of closely matching text.

Interview with Author A

JZUS editor: How did you come across Paper B?

Author A: I was updating my CV at the time and one of the pieces of information required was the citation of my papers. So I googled my papers and began compiling a list of papers that cited them. When I was doing that, I noticed that Paper B had a very similar title and abstract to mine (Paper A) that it cited. Upon further reading of the paper, I was surprised by how similar the main contents were too. Even worse was the fact that my name was misspelled in the reference list, which was really sloppy on the author's and the journal editor's part. After carefully comparing the two papers and consulting articles on plagiarism, I came to the conclusion that I had a case for complaint. Although the problem addressed by Author B was more specific, and he had made some small extensions to the analysis, it was obvious that he had borrowed a lot of materials from my previous paper, but without giving it due credit.

[1]OSI: The overall similarity index (OSI) is the percentage of similarity between a submission and information existing in the iThenticate databases selected as search targets [5, 14].

[2]SMSI: The single match similarity index (SMSI) is the percentage of similarity from a single source between a submission and information existing in the iThenticate databases selected as search targets [5, 14].

JZUS editor: What is your evidence that Paper B copied your research work?

Author A: I had worked on the problem for quite some time for my Master's degree. I was very familiar with its contents. There was no doubt in my mind that misconduct had been committed by Author B for the following reasons:

1. In the Abstracts, the beginning and concluding sentences were very similar between the two papers, even though they had been reworded in Paper B. These gave an inkling of what was to come.
2. Paper B had the same sections with the same titles, which were arranged in the same order as those in my paper. Within these sections, you could find 17 equations that were similar or even identical to those in my paper, together with many similar sentences, albeit reworded, to convey the ideas, assumptions and deductions. Paper B used these contents liberally without giving immediate citations to my paper; my paper was only cited in the final third of the paper when we came to two core equations in the derivation of the key parameter, the working of which was probably too long and tedious for Author B to reproduce in its entirety. Even then, the citation was no more than a mention in passing. According to my understanding, paraphrasing without giving credit is plagiarism, irrespective of whether the source is cited elsewhere in the paper. While he might not have done it deliberately, Author B was misleading the reader to think that all the foregoing equations and sentences were derived and composed independently by him.
3. More tellingly, for one important part of the analysis in my paper, I provided a reference to support the simplifying assumption that I had made for the boundary condition, as requested by one of the reviewers. It was intriguing to find that the same assumption with the same supporting reference also appeared in Paper B, but without mentioning my paper. It was highly unlikely that Author B came up independently with the same assumption which was supported by the same reference. That would have been an amazing coincidence. I consider this to be unequivocal evidence that Paper B took ideas and materials from my paper without acknowledging its source.

JZUS editor: What actions did you then take?

Author A: I first contacted the editor of Journal A that had published my paper, but he just told me to write to the other journal that published Paper B. When I did so, the reply I got from the editor (Editor B) was: "Shouldn't you contact the authors of the paper in question?" I then had to point out to him that, according to his own publisher's guidelines, it was his responsibility as the editor-in-chief to evaluate all cases of complaints and then to take appropriate actions where necessary.

JZUS editor: Do you feel that Editor B was avoiding his responsibility?

Author A: Yes, but I understand that dealing with alleged cases of plagiarism could be a time-consuming and troublesome affair, and many editors would prefer not to have to deal with them.

JZUS editor: What happened next?

Author A: A few days later, I got another reply from Editor B saying that he had reviewed both papers, but he did not see any problem as my paper was already cited by Author B. He also commented that the timing of my complaint was unusual, as both papers were published quite some time ago. I countered that, had he read my supporting document carefully, he would have realized that the very first citation of my paper only appeared more than half-way through Paper B, yet most of the materials that were presented before that point were taken from my paper. I pointed out to him that paraphrasing without properly and adequately acknowledging the source is indeed plagiarism. I also reminded him that search engines such as Google were recent inventions, and prior to their introduction, plagiarism was much harder to detect. People may think that I was making the complaint for my personal gain; however, I was already a tenured faculty member with an endowed position, working in a completely different field from my Master's topic, and I did not need the recognition for something that I no longer worked on. But I felt strongly that researchers such as Author B needed to be better educated on what constituted plagiarism. After all, is that not what we are teaching our undergraduate students?

JZUS editor: What was Editor B's reply?

Author A: He said he could see my point, but said that cases such as that of Paper B happened frequently, especially with authors whose mother tongue was not English. He also added that Paper B was not the type of case that he would 'go after'. It was then that I contacted you for help with CrossCheck, hoping that an independent check would help my case. Although the report from CrossCheck was not conclusive, your support and understanding kept me going. Shortly after that, I received an official email from the publishing manager of Journal B, who I guess was more familiar with the definition of plagiarism. He advised Author B and myself to discuss the matter in order to reach a satisfactory conclusion, and said that he and the editor anticipated that a suitable outcome would be a written note published in Journal B to clarify the full extent to which the findings presented in my paper should be acknowledged.

JZUS editor: That's encouraging; so did you contact Author B?

Author A: Yes, I did. I sent him, in my view, a very courteous email, saying that I did not believe that he had plagiarized intentionally but I would appreciate it if he were to publish a note acknowledging his oversight in not giving my work adequate citation.

JZUS editor: How did he take it?

Author A: Initially, he was very apologetic (for spelling my name wrongly in the reference), adding that my paper had been very useful for his work. However, when it became clear to him that my complaint was of a more serious nature, he became defiant. He argued that the problem he had addressed was so different from mine—actually, it was only more specific—that he did not feel the need to cite my paper earlier, even though he had admitted that his work was built on mine and papers dealing with general problems similar to the one in mine were quoted in his introduction.

JZUS editor: So did you come to an agreement?

Author A: It was clear from our communications that Author B's English was not that great. Finally, after several rounds of email communication that did not go anywhere, I got so frustrated that I took the liberty of drafting a note for him which basically stated that he should have acknowledged my work earlier and more adequately in his paper. I must emphasize that the word plagiarism was not mentioned in the note at all. Throughout, to keep them informed, I copied all my communications with Author B to Editor B and the publishing manager. However, Editor B did not like my note at all; he sent me a very stern email saying that I had crossed the line. He said that it was solely up to Author B to decide whether to submit a note or not, and he, as the editor-in-chief, had the right to decide whether or not to publish the note, even when it was submitted to the journal. He even questioned my motive for making the complaint.

JZUS editor: That must have been very frustrating for you; did you give up at this point?

Author A: No, I did not. I told Editor B that, referring again to the web site of his journal's Publisher, it was really his responsibility to consider and evaluate the responses and make a judgment, based on his knowledge and experience in the field and then take appropriate action. The guidelines clearly stated that these matters should not be abandoned. It later dawned on me that if Author B had taken material from my work, he could have done the same with other people's work, too. Lo and behold, I soon found that he also "borrowed" a lot of materials from another paper, again without giving due credit to the authors. I duly presented my new findings to Editor B and the publishing manager, adding that if Author B refused to send in a note acknowledging his mistakes (the word I used, to save his face, was oversights), I would submit an article summarizing my findings on the two papers and let people make up their own minds. It did not take long before the publishing manager wrote back to say that they would consider my submission as Author B would not agree to publishing the requested note, but he insisted that my article would only be published if the editor considered it beneficial to their readers.

JZUS editor: Did you go on to submit your article?

Author A: I have to say that, by that time, I had been so upset by Editor B's responses that I was determined to prove him wrong. However, when I was about to submit my article, I decided to contact Author B's mentor who was also a co-author on the paper. He was a well known researcher and I wanted him to plead with Author B to come to his senses, as I realized that if the article did get published, his reputation would also suffer.

JZUS editor: What did the mentor say?

Author A: He confided in me that he had had many sleepless nights since learning about my complaint. But he had no doubt about Author B's integrity. He said he was a hardworking individual but his English was poor, perhaps far from sufficient to express himself accurately. Like Editor B, he did not think the lack of adequate citation was a serious problem; he certainly did not think it was plagiarism. He went on to say that some of his works had also been used by others without citation, but he looked at the matter from a positive viewpoint, and took it as a compliment when his work was used, whether or not it had been cited.

JZUS editor: Did you feel that Author B's mentor also did not fully understand the definition of plagiarism?

Author A: That seems to be the case, and I believe that this problem is actually quite common even amongst established engineering and physical sciences researchers. However, it is simply not right to fail to cite people's work where credit is due. It is very unfair to the researchers who have done all the hard work. Having problems with the language cannot be used as an excuse.

JZUS editor: Was your article published?

Author A: No. I never submitted it. I had calmed down by then. I did not want to cause irrevocable damage to anyone's career or reputation. I also realized that the whole thing had become a huge distraction for me; I could not focus on other, more important jobs. I therefore decided to move on. But I still wanted people to have a better understanding of what plagiarism is by learning from Author B's mistakes, so that similar things would not happen again. I also wanted editors to take a more balanced view and a more proactive role in dealing with cases of alleged plagiarism. That's why I approached you with the proposal of publishing my story anonymously. Given the emphasis you place on countering plagiarism and improving the overall quality of papers by Chinese researchers, I think *JZUS* would be the ideal place for my story to be told.

Finally, *JZUS* editor asked Author A whether, if he had been the reviewer of Paper B, he would have rejected it outright. Much to our surprise, he said he would not. He went on to explain that provided his work had been given enough credit and cited properly, he would recommend Paper B be accepted for publication after some revisions, since it did extend his own work. It is clear that Author A is a decent and fair scientist, but nevertheless he would not tolerate plagiarism.

Discussion

Today, not only do we know 'Plagiarism is more than theft. It represents a challenge to your integrity and expertise and puts your reputation on the line' [21], but are aware of 'Not so many years ago, we got one or two alleged cases a year.

Now we are getting one or two a month' [22]. There is an increasing challenge to find and deal with plagiarism. However, everyone has a common duty to clean the academic environment.

Problem and Limitation

The case study presented above highlights the inadequacy of purely text-matching tools such as CrossCheck for identifying plagiarism in papers in fields such as engineering and physical sciences since such tools are unable effectively to compare the formulae, tables and figures that frequently form the bulk of the content of the papers. Unfortunately, as far as we are aware, no suitable tool currently exists to spare editors and peer reviewers the arduous task of comparing non-textual elements in papers where plagiarism may be suspected.

The case study also highlights the difficulty of detecting plagiarism that involves paraphrasing with no or inadequate citation. As discussed in the ACM Policy and Procedures on [23] 'plagiarism manifests itself in a variety of forms, also including verbatim copying of portions of another author's paper with citing but not clearly differentiating what text has been copied (e.g., not applying quotation marks correctly) and/or not citing the source correctly'. Nevertheless, 'if the case can be made that one's work consists predominantly of someone else's words or ideas, one may still be susceptible to charges of plagiarism' [10, 24]. It is often difficult for editors to determine whether authors have obeyed the rules.

As journal editors, we cannot help but ask, what to do with this kind of problem?

Experience and Responsibility

For users of CrossCheck in our experience [19], a lower SMSI may be appropriate to highlight papers containing possible plagiarism for which we can refer to CrossCheck_Manual [7] and IEEE CrossCheck user's guide [25]. The editors should then compare the papers returned with the suspect paper, paying special attention to similar contents that contain no citations, including the captions and main-body text that describe figures, tables and formulae. There are warning signs, too. Because plagiarized material is written for other purposes, it is often slightly off topic, has odd references or content, unusual phrasing or vocabulary, and maybe the level of the work could not match the known standard performance of the researcher [8].

If the journal editor lacks expert knowledge in the specific topic of the paper, he or she may be unable to distinguish between generally accepted facts, arguments and results and those given or derived by specific researchers, warranting proper citation. This is just one reason why high-level peer review should always be carried out by experts in the subject (who, of course, do not have any conflict of interest due to current or previous collaboration or other relationship with the authors of the paper in question) [11, 26]. The journal's guidelines for reviewers should always include a request to determine whether the paper appears to contain elements of plagiarism and, if so, to indicate whether in their opinion the paper should be rejected outright or should be revised to include the appropriate citations.

This case also touches on the difficulties of finding ideas plagiarism. As Author A said in our interview that, "Paper B took ideas and materials from my paper without acknowledging its source", he thought that if Editor B could consider him as a reviewer at the beginning, this event will be avoided. This implies that it is very important and is the responsibility for the journal editor to find appropriate referees in this area, specially paying attention to those authors whose papers are cited in the key parts of the current article. In our editorial experience, maybe the appropriate reviewer should be the best hunter to judge idea plagiarism.

Nowadays journal editors not only do their normal editing and publishing work, but also spend much time to deal with misconduct problems [18, 19, 27, 28]. However, it is not the excuse to escape the responsibility to deal with the misconduct and solve the dispute. In this case study, similarly, the dispute could have been resolved had Editor B taken a more proactive role in the discussion between the authors in order to reach a satisfactory conclusion for all concerned; leaving the authors to sort out their differences clearly did not help. The COPE Code of Conduct for editors stated that 'Editors should be responsible for everything published in their journals', and 'always be willing to publish corrections, clarifications, retractions and apologies when needed' [29, 30].

Ethic Education's Necessity

Our case study also identifies a lack of understanding on what constitutes plagiarism, both by a seasoned researcher and, rather worryingly, by the editor of an international journal. Author A, through his detailed investigation and comparison between his and Author B's papers, demonstrated unequivocally that plagiarism—specifically paraphrasing without prompt and proper citations—had been committed by Author B, even if unintentionally. It is surprising that, when presented with these findings and reference to the guidelines given by the very publisher of the journal involved, both the senior author of the paper in question and the editor-in-chief did not appear to believe that any serious wrongdoing had been committed by Author B.

While online guides and tutorials are available for students and researchers in many universities [31], familiarization with the relevant rules should be made compulsory as part of their training and given special emphasis. Authors should be properly educated in what constitutes plagiarism and how to avoid it by giving adequate credit to the originator of the ideas or text that they copy or paraphrase. Researchers who are not native English speakers will then no longer be able to use deficient language skills as an excuse for their "oversights".

Watson and Hayter [29] commented that for sometimes it is difficult to find obscure misconduct, such as plagiarism etc., in the process of peer review, some organizations, like universities, should vet research outputs before they get to the publishing (and scandal) stage. Research institutions have the responsibilities to investigate misconduct, correct the scientific literature, and furthermore, prevent misconduct and its consequences [27].

In a short, authors, journal editors, reviewers, as well as research institutions all have responsibilities not only to train themselves, but also to set criteria so that the next generation can recognize the differences between ethical and unethical behavior [32].

Common International Publishing Ethical Standards

In recent years, some handbooks of journal publishing or essays writing for editors and authors have presented references to international publishing ethical standards [11, 26, 33]. The first issue of *Nature* in 2012 showed a special comment to discuss "how to stop plagiarism", which applied some active advices [9]. Furthermore, the COPE gave helpful standards, guidelines, and many cases about publication ethics, which are widely accepted by many publishers, e.g., the international standards for editors and authors, the flowchart on how editors should respond to suspected plagiarism, guidelines for journal editors, the cooperation between research institutions and journals on research integrity cases, as well as the retraction guidelines when plagiarism is confirmed, etc. (http://publicationethics.org/resources).

Conclusions

Editors should apply a lower threshold when using detectors, such as CrossCheck to identify potential plagiarism, especially papers are heavy in non-textual content, such as engineering papers. It goes without saying that all papers should be reviewed by relevant experts who have no connection with the authors; guidelines to reviewers should include the need to identify potential plagiarism and to recommend appropriate action. If plagiarism is confirmed, editors and publishers should have the courage and responsibility to take the appropriate action. The COPE (http://publicationethics.org/) has given helpful and valuable standards, guidelines, cases, etc. about publication ethics for authors, journal editors and publishers, which are widely accepted.

Everyone has a responsibility to promote a culture in which research misconduct does not happen [27]. Only then, can plagiarism be eliminated and the integrity of scientific research be protected.

Practical Lessons for Authors and Editors

The case shows that some authors or editors themselves have a lack of understanding of plagiarism. It also illustrates the difficulty of getting some editors and publishers to take appropriate action when plagiarism is identified. This makes it all the more important that authors, journal editors, and reviewers, as well as research institutions, all take responsibility not only for recognizing the difference between ethical and unethical behaviors, in order to protect a healthy research environment, but also for maintaining consistent ethical publication standards.

Guidance for Authors and Editors If an author reuses another author's equations, images, tables or data, or their actual ideas, without acknowledgment of the original source together with a full citation, this is considered plagiarism.

Authors should always identify and acknowledge the originator of the material or idea(s), and provide a full citation to the original published source.

Editors should be careful to choose appropriate reviewers. It is often a good idea to select reviewers from among the authors cited in the article, because they are most likely to be very familiar with their own and others' work on the subject of the article. If this type of plagiarism is detected after the publication and an agreed solution cannot be reached, editors should refer the issue to an expert body, such as COPE [34].

References

1. Zhang, X.X., Z.L. Huo, and Y.H. Zhang. 2014. Detecting and (not) dealing with plagiarism in an engineering paper: beyond CrossCheck—A case study. *Science and Engineering Ethics* 20(2): 433–443. doi:10.1007/s11948-013-9460-5.2014. (with permission of Springer).
2. Noorden, R.V. 2015. The image detective who roots out manuscript flaws. *Nature*. doi:10.1038/nature.2015.17749. http://www.nature.com/news/the-image-detective-who-roots-out-manuscript-flaws-1.17749.
3. ORI: For newer versions of Photoshop, please go to: http://ori.hhs.gov/advanced-forensic-actions.
4. Garner, H.R. 2011. Combating unethical publications with plagiarism detection services. *Urology and Oncology* 29(1): 95–99.
5. Wager, E. 2011. How should editors respond to plagiarism? Available from http://publicationethics.org/files/Discussion%20document.pdf. Accessed 11 August, 2012.
6. Baždarić, K., L. Bilić-Zulle, G. Brumini, and M. Petrovečki. 2012. Prevalence of plagiarism in recent submissions to the Croatian Medical Journal. *Science Engineering Ethics* 18: 223–239. doi:10.1007/s11948-011-9347-2.
7. CrossCheck_Manual. 2009. http://www.crossref.org/crosscheck/CrossCheck_Manual.pdf. Accessed September 4, 2012.
8. Abdelmoneim, S.E. 2010. Plagiarism: What is it? How to avoid it? 14th Alexandria Anaesthesia & Intensive Care Conference. Alexandria Faculty of Medicine. Available from http://www.alexaic.com/alexaicfiles/presentation2010/day3/028001.pdf. Accessed August 1, 2012.
9. Garner, H., B. Pulverer, A. Marušić, M. Petrovečki, J. Loadsman, and Y.H. Zhang. 2012. Comment: How to stop plagiarism. *Nature* 481: 21–23.

10. Bouville, M. 2008. Plagiarism: Words and ideas. *Science Engineering Ethics* 14: 311–322. doi:10.1007/s11948-008-9057-6.
11. Morris, S., E. Barnas, D. La Frenier, and M. Reich. 2013. *The handbook of journal publishing*. Cambridge: Cambridge University Press.
12. Brinkman, B. 2013. An analysis of student privacy rights in the use of plagiarism detection systems. *Science Engineering Ethics* 19: 1255–1266.
13. COPE. 2010. CrossCheck guidance: An analysis of typical cases of plagiarism in different disciplines. Available from http://publicationethics.org/resources/research. Accessed on July 13, 2013.
14. Zhang, Y.H. 2010. CrossCheck: An effective tool for detecting plagiarism. *Learned Publishing* 23: 9–14. doi:10.1087/20100103.
15. Zhang, Y.H. 2010. Chinese journal finds 31 % of submissions plagiarized. *Nature* 467(9): 153.
16. Zhang, Y.H., and X.Y. Jia. 2012. A survey on the use of CrossCheck for detecting plagiarism in journal articles. *Learned Publishing* 25(4): 292–307.
17. Jia, X.Y., X.F. Tan, and Y.H. Zhang. 2013. Replication of the methods section in biosciences papers: Is it plagiarism? *Scientometrics*. doi:10.1007/s11192-013-1033-5.
18. Zhang, Y.H., and X.Y. Jia. 2013. Republication of conference papers in journals? *Learned Publishing* 26(3): 189–196.
19. Zhang, Y.H., X.Y. Jia, H.F. Lin, and X.F. Tan. 2013. Editorial: Be careful! Avoiding duplication: A case study. *Journal of Zhejiang University-SCIENCE B (Biomedicine & Biotechnology)* 14(4): 355–358. doi:10.1631/jzus.B1300078.
20. Li, Y.Y. 2012. Text-based plagiarism in scientific publishing: Issues, developments and education. *Science Engineering Ethics* 19(2): 569–583. doi:10.1007/s11948-012-9367-6.
21. Bugeja, M. 2012. Busting the new breed of plagiarist. http://www.awpwriter.org/magazine/writers/mbugeja01.htm. Accessed July 5, 2013.
22. Butler, D. 2010. Journals step up plagiarism policing cut-and-paste culture tackled by CrossCheck software. *Nature* 466: 167. doi:10.1038/466167a.
23. ACM Policy and Procedures on Plagiarism. first issued in Oct 2006, revised June 2010. (http://www.acm.org/publications/policies/plagiarism_policy).
24. Plagiarism FAQs. 2013. Available from http://plagiarism.org/ask-the-experts/faq. Accessed April 10, 2013.
25. IEEE CrossCheck user's guide. 2012. Available from http://www.ieee.org/documents/CrossCheck_User_Guide_S1M119068.pdf. Accessed September 4, 2012.
26. Hames, I. 2007. *Peer review and manuscript management in scientific journals: Guidelines for good practice*, 1st ed. Oxford: Wiley.
27. Sox, H.C., and D. Rennie. 2006. Research misconduct, retraction, and cleansing the medical literature. Lessons from the Poehlman case. *Annuals of Internal Medicine* 144: 609–613.
28. Wager, E., and P. Williams. 2011. Why and how do journals retract articles? An analysis of Medline retractions 1988–2008. *Journal of Medical Ethics* 37: 567–570.
29. Watson, R., and M. Hayter. 2013. Halt fraud before it hits the headlines. Times Higher Education, April 4, 2013. Available from http://www.timeshighereducation.co.uk/comment/opinion/halt-fraud-beforeit-hits-the-headlines/2002887.article. Accessed April 15, 2013.
30. COPE. 2011. Code of conduct and best practice guidelines for journal editors. Available from http://publicationethics.org/files/Code_of_conduct_for_journal_editors.pdf. Accessed July 4, 2012.
31. Maurer, H., F. Kappe, and B. Zaka. 2006. Plagiarism—A survey. *Journal of Universal Computer Science* 12(8): 1050–1084.
32. Carraway, L.N. 2009. Ethics for and responsibilities of authors, reviewers and editors in science. *The American Midland Naturalist* 161: 146–164.
33. Greetham, B. 2008. *How to write better essays*, second edition. Basingstoke, England: Palgrave.
34. COPE: Responsible research publication: international standards for editors. http://publicationethics.org/files/International%20standard_editors_for%20website_11_Nov_2011.pdf.

Part II
Discipline-Specific Plagiarism Issues

Chapter 5
Biosciences: Replication of Methods Sections

Focus on Subject Differences

In our first survey (see Chap. 2), we noticed some significant differences in the responses to some of the questions between the subject areas of Biomedicine and Biotechnology, and Computers and Electronics. We therefore decided to conduct a further investigation focusing on these key differences in approach in these disciplinary fields.

In conducting CrossChecks on our own journals we had frequently noticed that replication of methods sections was particularly common in bioscience papers, whereas republication of conference papers was most common in computing journals. We felt that it would be useful to investigate journal editors' attitudes to these special problems in their areas: what they considered acceptable, and how they approached the issue.

With the support of a grant from the COPE research fund, we therefore carried out three surveys as follows:

1. Survey on author's attitudes to copying in Methods Section of a Paper (February 2012)
2. How do you handle the Methods Section of your paper in the Life Sciences field? (July 2012)
3. How do journal editors handle the re-publication of conference proceedings papers in Computing or Electrical and Electronic Engineering fields? (July 2012)

Y.H. Zhang, *Against Plagiarism*, Qualitative and Quantitative Analysis of Scientific and Scholarly Communication,
DOI 10.1007/978-3-319-24160-9_5

We published the findings in a number of journal articles [1–3] which are reproduced in their entirety and discussed in detail in this chapter and Chap. 6.

Replication of Methods Sections: What Is the Problem?

In bioscience papers, replication of the methods section is a common problem. In our own experience as editors, CrossCheck has frequently detected duplication in the Materials and Methods sections. In addition, we often encounter authors who believe that it is acceptable to copy methods sections from their own published materials, as opposed to those of others. In this chapter we address how we think editors should handle papers which include duplicated content in the method sections, and provide guidance for authors on how to write their methods sections without committing plagiarism.

The text that follows reproduces, with the publisher's permission, our article 'Replication of the methods section in biosciences papers: is it plagiarism? [1], which reports the findings of a study that we conducted to investigate this issue. Firstly, we investigated the behavior of authors (who were also experienced reviewers) when writing the methods section in their published papers. Then we looked at how the top journals in this area deal with the problem, by taking the descriptions of one well-established method in randomly selected papers published in eight top journals, and analyzing them using CrossCheck to identify the extent of duplication. The survey results show that an author may employ one of a number of approaches to writing the methods section within a paper; repeating a previously published method is more common than either giving a citation only, or totally rewriting the description in the author's own words. In the samples from the eight leading journals, we detected very little plagiarism. For common methods, the journal Science tended to include the description as a separate online attachment. We also reproduce, with the publisher's permission, a case study [2] on the same topic.

Introduction

Verbatim repetition of another author's words or of the author's own words, without proper acknowledgement, constitutes plagiarism or self-plagiarism (redundant publication) [4]. Even when acknowledged, a high proportion of quotation in an article tends to suggest lack of originality. However, the methods sections of research articles pose a particular problem. Roig proposes "teaching scientists to paraphrase" to reduce plagiarism in science publishing [5]; however, in the ensuing correspondence commented that "provided it is properly attributed, the repetition of the standard description of experimental procedures is necessary—even though it is likely to be detected by plagiarism detection software—because each instance of paraphrase could lead to a 'procedure drift', causing variations from the original procedures." Obviously, Roig did not omit this point. He had expressed the same argument only under the assumption of the nearly perfect methods description that any additional clarification or elaboration is often unwarranted [6].

COPE flowcharts demonstrated that an editor's response to plagiarism should depend on the type and extent of the copying [7]. Since the authors started using CrossCheck to detect possible plagiarism in their journal, they have frequently discovered repetition in the methods sections of bioscience papers [8]. In an earlier study by the present authors investigated the attitudes of journal editors towards duplication in the methods section [9]. The results showed that editors generally found it unacceptable to duplicate the description of methods, unless it was either rewritten using the author's own words or properly cited. However, 20 % of journal editors in the biosciences considered it acceptable to include between 20 and 40 % of duplicated content in the methods section. This gave rise to two questions: (1) Does duplication in the methods section constitute a form of plagiarism? (2) Is it permissible in the methods section?

In the present study, firstly, the authors surveyed the behavior of authors when writing the methods section in their published papers. Secondly, the descriptions of one well-established method in randomly selected bioscience papers published in top journals were analyzed using CrossCheck to identify the extent of duplication. Finally, we give our own recommendations for preparing the methods sections for biosciences papers based on the above analysis.

What Authors Do?

Survey Design

An online survey was created, consisting of five questions on the respondents' language (Q1), publishing experience (Q2), their behaviors in writing the methods section in their three recently published papers (Q3 and Q4), and their experience as reviewers (Q5).

An email invitation to participate was sent to a total of 2177 recipients chosen from the authors' journal's own database of international reviewers in the field of biosciences (http://www.zju.edu.cn/jzus/reviewer.php#B) [10, 11]. The reviewers in this database are authors from all around the world, who have published in international journals. Respondents were directed to the survey at SurveyMonkey (https://www.surveymonkey.com) from July 9, 2012 to July 31, 2012.

Survey Results Analysis

In the survey, respondents were asked to provide details of their 3 most recently published papers (Q3) and to indicate their writing behaviors in the methods section of these papers (Q4)—thus the answers to each part of Q4 should have been 0, 1, 2 or 3. However, a few respondents misunderstood the question, and gave numbers larger than 3. Such responses were therefore filtered out, leaving 178 valid responses out of the original 192.

Of the 178 respondents, 101 (57 %) were native English speakers (Q1) while the remains were using other language. According to their answers to Q2, 81 % (145/178) had published more than 20 papers, while only 6 % (10/178) had published fewer than 10 papers.

Details of the three most recently published papers from the 178 respondents are as shown in Table 5.1 (Q3). The 2011 impact factors of these journals were obtained by searching the database of Journal Citation Reports [12]. Seven papers were published in six leading journals with impact factors (IFs) > 30: *The New England Journal of Medicine*, *The Lancet*, *Cell*, *Science*, *Nature*, and *The Journal of the American Medical Association* (*JAMA*); four papers were published in four journals with IFs of between 20 and 30. The most frequently mentioned journal was *PLoS One*, with 22 papers.

Table 5.1 Journal information of the three most recently published papers from the 178 respondents (Q3)

2011 IF	No. of papers	No. of journals	Journal title (No. of articles)
>30	7	6	N. Engl. J. Med. (1), Lancet (1), Cell (1), Science (2), Nature (1), JAMA (1)
20–30	4	4	Cancer Cell (1), Nature Immunol. (1), Lancet Oncol. (1), Nature Med. (1)
10–20	22	17	Nature Cell Biol. (1), J. Clin. Oncol. (2), Ecol. Lett. (1), PLoS Med. (1), Nature Neurosci. (1), Circulation (1), Mol. Cell (1), J. Am. Coll. Cardiol. (3), Adv. Mater. (1), J. Clin. Invest. (2), Nature Struct. Mol. Biol. (1), Gastroenterology (1), Hepatology (1), Genes Devel. (2), Ann. Neurol. (1), Trends Mol. Med. (1), J. Cell Boil. (1)
5–10	95	71	PNAS (4), Arterioscler Thromb. Vasc. Biol. (4), Plant Physiol. (4), J. Neurosci. (3), Anal. Chem. (3), J. Immunol. (3), other journals not listed
3–5	184	121	PLoS One (22), J. Biol. Chem. (6), Other journals not listed
<3	196	159	Journals not listed
None	26	26	Proceedings or journals not indexed in JCR or science citation index
Total	534	404	Above

From the above information, we can infer that most of the respondents are experienced authors. Their indicated behaviors when writing the methods section of their listed papers (Q4) are as shown in Fig. 5.1.

In Q4, we classified these behaviors broadly into the categories: repetition (eight forms), provision of citation only, and complete rewriting in the author's own words. Repetition was subdivided according to the nature of the original source (own or others'), whether or not the text had been reworded, and whether or not a citation to the original source was provided. More than half (53.4 %) of the respondents reported that they utilized more than three possible behaviors in their three papers; this suggests that an author may use more than one style in the methods section of a single paper (e.g. when describing different methods).

As Fig. 5.1 shows, in most cases (446/535, 82.6 %) authors provide a citation to the previously published source of the description, whether written by themselves or others. The three most commonly reported behaviors are:

1. Completely use one's own words (294 cases, 31.1 %);
2. Repeat one's own published methods with some rewording and citation (216 cases, 22.9 %);
3. Repeat others' published methods with some rewording and citation (116 cases, 12.3 %).

Some form of repetition (535 cases) was more common than complete rewriting using one's own words (294 cases) or providing a citation only (115 cases). This is easy to understand, as many bioscience methods are frequently used in a variety of different studies.

In the 131 cases, which indicated that they had repeated one or more descriptions verbatim, 84 % (110/131) of these gave a citation and 16 % (21/131) did not (13.7 % in reusing their own description, and 2.3 % in reusing another author's description). Verbatim repetition was less common with others' work than with one's own.

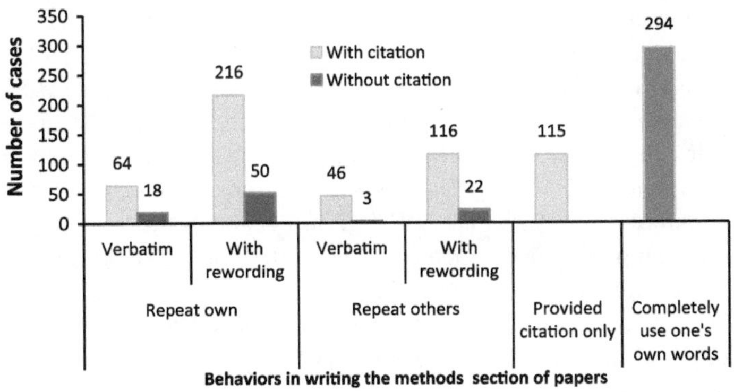

Fig. 5.1 Author behaviors in the methods sections of the above listed three most recently published papers (Q4)

In 404 cases the description had been repeated with some rewording. 82 % of these provided a citation and 18 % did not (12.4 % in reusing their own description and 5.4 % in using another author's description).

From the above analysis, it is clear that an author may employ different approaches to writing the methods section within a single paper. Authors are more likely to repeat the description of a method than simply to provide a citation. When they do repeat a published description, authors are more often to take it from a previously published paper of their own than someone else's; authors are more likely to make some modifications than to copy verbatim; and in the majority of cases they do provide a citation to the previously published source. Authors are more likely to provide a citation when repeating the description of a method verbatim, than when they have done some rewording.

What Top Journals Do?

To ascertain the extent to which existing published descriptions were replicated, with or without citation, we examined the practice as exemplified in the treatment of one particular well-established method in a sample of articles from eight leading journals.

Methods

The increasingly widespread use of CrossCheck, in recent years, to detect potential plagiarism has highlighted the extent of textual duplication in the methods sections of journal papers, particularly in the biosciences [6, 13]. To establish best practice in the field, the authors selected eight leading bioscience journals and surveyed their articles.

The journals selected, mostly from among those listed by respondents to the first survey, were those with the highest impact factors in the field, from which the full text of relatively recently published articles was available to the authors for analysis (in some cases the PubMed database did not contain the most recent articles). These criteria led to the selection of the following journals: *Cell*, *The Journal of the American Medical Association*, *The Lancet*, *Nature*, *Nature Biotechnology*, *The New England Journal of Medicine*, *PLoS Biology*, and *Science*.

One most commonly used methods, 'western blot' (also known as western blotting, electrophoretic transfer of proteins from sodium dodecyl sulfate-polyacrylamide gels to unmodified nitrocellulose and radiographic detection with antibody and radioiodinated protein A [14]), was selected, and articles in the eight selected journals were identified in PubMed by searching the PubMed database using the term 'western blot*'. This search produced a total of 512 articles. For each journal, the 10 most recently published available articles were downloaded and the methods sections were analyzed.

The methods sections were classified as falling into one of the following categories: 'description', 'citation' and 'attachment'. For those articles which fell into the 'description' and 'attachment' categories, CrossCheck was run on the relevant part of the methods section to ascertain whether previously published text was replicated, with or without citation.

Analysis of Methods Sections in the 10 Most Recent Available Articles from Each Journal

Three types of expression were found for the methods section relating to western blot. These were classified as 'description' (fully described the procedures used), 'citation' (simply provided a citation to a previously published description or website), and 'attachment' (supplied the detailed description as an attachment, rather than in the main text).

All three types of expression were found in *the Lancet* and *the New England Journal of Medicine,* but only two out of three in the six remaining journals; *Science* used the 'attachment' approach in all but one case. Table 5.2 shows the results of this analysis.

Table 5.2 Analysis of expression styles used in methods section

Journal title	Total articles	No. articles utilizing each style		
		Description	Citation	Attachment
Cell	10	8	2	0
The journal of the American medical association	10	9	1	0
Nature	10	8	2	0
Nature biotechnology	10	8	2	0
PLoS biology	10	7	3	0
Science	10	1	0	9
The lancet	10	7	2	1
The New England journal of medicine	10	7	1	2

CrossCheck Results

1. **'Description'**

This was by far the most frequently adopted style of the methods section for all of the journals studied, apart from *Science*. For the 55 articles containing a 'description', a CrossCheck was run on the relevant part of the methods section. Three articles—1 article in *Lancet*, 1 in *JAMA*, and 1 in *Nature Biotechnol.*—were found to contain paragraph matches, i.e. nearly no less than one paragraph matched with previously published materials. In the case of both *Lancet* and *Nature Biotechnol.*, the articles studied and the articles replicated were by one or more of the same authors (neither of the sources of the replicated text was cited where it appeared); in the case of *JAMA*, although there was no common author, the article replicated was cited elsewhere in the text; however, the replicated text was nearly 60 words long. No other articles, in any of the journals, contained any obviously replicated content in the relevant part of the methods section. Table 5.3 shows the results for the 'description' category of articles.

Table 5.3 CrossCheck results for articles using 'description'

Journal title	No. articles	No. articles with paragraph matches	Replicated text		Remark on replicated text
			With citation	Without citation	
Cell	8	0			
JAMA: the journal of the American medical association	9	1	1		Replicated a section of c. 60[a] words from previously published article (uncited but cited elsewhere in the article under study)
Nature	8	0			
Nature biotechnology	8	1		1	Replicated a section of c. 45 words from previously published article by same authors (uncited)
PLoS biology	7	0			
Science	1	0			
The lancet	7	1		1	Replicated a section of c 45 words from previously published article by same authors (uncited)
The New England journal of medicine	7	0			

[a]As CrossCheck could not distinguish some symbols like "μ" "χ" and so on, the number of repeated words is a approximate number

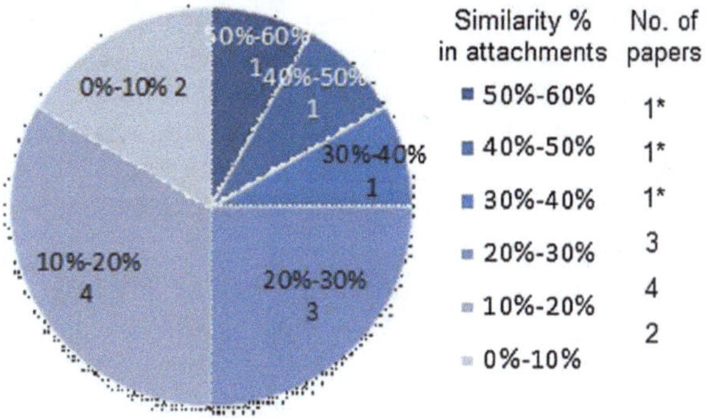

Fig. 5.2 Similarity scores for attachments (supplemental materials). *Asterisk* Attachment with high similarity score in *Science*

2. **'Attachment'**

For the articles in *Science*, 'attachment' (the inclusion of supplemental materials describing the method in detail) was by far the most frequent style of expression in methods sections; 9 of the 10 articles searched (all published since 2002) were found to use this approach. By contrast, just one article in *Lancet* and two in *N. Engl. J. Med.* used the same approach, while it did not occur at all in the sample articles from the other journals studied. A CrossCheck was carried out on the text included in the 12 attachments.

Of these 12 articles, 3 (all in *Science*) were shown to include paragraph matches—entire sections of replicated text from elsewhere. In two cases, the original source of the replicated text was not indicated. In one case, although there was no common author, the article replicated was cited elsewhere in the text; however, the source of the replicated text was not indicated where it appeared.

Figure 5.2 shows the remarkably high similarity scores [15] that resulted from this check. In three articles out of the nine, the similarity score was above 30 %—in one it was above 50 %. In some cases, the replicated material was from articles by other authors, and the source was cited elsewhere in the article, although it was not indicated where it appeared; in others it was from previously published articles by the same authors, but without citation. It would appear that, in *Science* at least, authors and editors are less concerned about replication of text in supplemental materials than in the main text.

Results Analysis

The study showed that the eight leading journals differed in their handling of the description of a well-established method. While description in the main text was the most common approach (55 articles out of 80), and citation only was the next most common (13 out of 80), attachment of the description as supplemental material (12 out of 80 overall) was the strongly preferred approach in *Science* (9 out of 10 articles).

CrossChecks demonstrated a relatively low incidence of textual replication where the 'description' approach was adopted (although citation practice was poor in these cases). However, where the 'attachment' approach was adopted (mainly in *Science*), the level of textual replication was extremely high, and the original sources were rarely indicated.

With either the 'description' or the 'attachment' approach, citation practice was extremely poor.

From the study of the eight leading journals, plagiarize is very rare in such journals. And learning from *Science*, attachment may be a considerable choice for papers with common methods. However, whether high similarity scores in attachments are acceptable, a conclusion could not be drawn yet. Further researches still need to be done.

Discussion and Suggestions

In the field of biosciences, most journals require that a research article has a section named "Materials and Methods" (or a similar section) to show how the work is carried out. Kallet proposed that the methods section should provide sufficient information to enable readers to judge the validity of a study: "a clear and precise description of how an experiment was done, and the rationale for specific experimental procedures, are crucial aspects of scientific writing" [16]. To tutoring authors in preparing the manuscript, some journals (e.g., *Genome Biology*, *The EMBO Journal*, *PLoS Genetics*, *Journal of Endocrinology*) clearly stated the requirement to the Materials and Methods section for research articles in its Instruction for Authors.

As shown by the survey results, authors can express the methods to readers in different ways in one paper. This may attributed to the method itself, since well-established methods are familiar and easy for readers to understand, and new methods are usually need detailed description for readers to understand.

In the eight leading journals sampled in this paper, the preferred approach when describing a well-established method was to use the author's own words. If some or all of the description is replicated verbatim from a previously published article (whether by the same or different authors), appropriate acknowledgement and citation must be given to the original publication [17, 18], but this was not observed in any of the cases studied. Alternatively, unless the methods used are totally novel, a phrase such as 'by well-established method' or 'as previously described' plus appropriate citation may be used. However, the editors of *Science*, as stated in the journal's author guidelines (http://www.sciencemag.org/site/feature/contribinfo/prep/prep_online.xhtml), prefer the attachment of supplemental material (in the present authors' view, the requirement for acknowledgement and citation should be no less in the case of an attachment than in a description).

Based on the above discussion, we can give the following advice for authors preparing the methods section of bioscience papers:

1. Wherever possible, authors should use their own words to describe the method.
2. If all or part of a previously published description—whether by the same or different authors—is unavoidably repeated, the repeated text should be clearly identified (e.g. by indentation or quotation marks), and a full citation given to the original source. (A case study can be referred in the appendix.)
3. Different journals prefer different approaches. Some require the use of a phrase such as 'by well-established method' or 'as previously described', plus appropriate citation to an existing description for any well-established method, and only require a detailed description if the method used is totally novel. Others again require authors to attach a description of the method as part of the supplemental materials; in this case, the same rules should be followed as in point 2 above.

If these guidelines are followed, readers will be given enough information to evaluate the study described in the paper and, if necessary, to replicate it. At the same time, not only will tools such as CrossCheck not find high levels of similarity with the authors' own or others' work, but—more importantly—the authors will have avoided committing the unethical practices of plagiarism or self-plagiarism.

Avoiding Duplication: A Case Study

In the *Journals of Zhejiang University-SCIENCE (A/B/C)*, we have used CrossCheck since 2008, and we have found it to be a very effective tool for detecting duplicated content (thus suggesting the possibility of plagiarism). However, while CrossCheck is enables the editor to find strings of similar text, most instances of true plagiarism cannot be identified solely by these strings; editorial judgement is required.

To understand what the CrossCheck reports mean, it is necessary to understand two concepts: the Overall Similarity Index (OSI) and the Single Match Similarity Index (SMSI). The OSI gives the total percentage of similarity between a submission and information existing in the CrossCheck/iThenticate databases selected as search targets. The SMSI gives the percentage of similarity from a single source between the iThenticate database and the submitted document. If a paper has an OSI > $\sim$40 % or an SMSI > $\sim$10 %, we usually reject it out of hand; if the OSI is between $\sim$25 and $\sim$40 % or the SMSI is between $\sim$3 and $\sim$10 %, and if it is our editorial judgement that the ideas may have been plagiarized, the paper is usually returned for revision. Papers with an OSI < $\sim$25 % and SMSI < $\sim$3 %, are handled on a case-by-case basis; in most cases, plagiarism is unclear, and we have to send the CrossCheck similarity report to the author to clarify whether plagiarism is involved or not. If plagiarism is acknowledged, the editors usually either reject the article outright or return it for the author to re-write, depending on the particular circumstances. As CrossCheck can only detect duplication in text, and not in figures, graphs etc., (and neither, of course, can it detect the plagiarism of ideas,

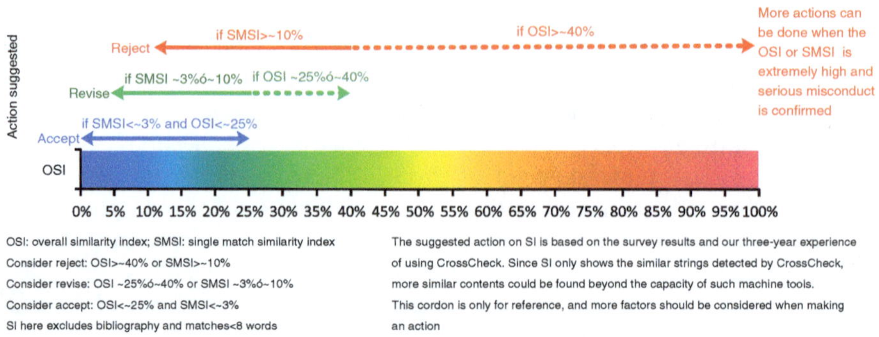

Fig. 5.3 Similarity score cordon of *Journals of Zhejiang University-SCIENCE (A/B/C)* for CrossCheck user reference

as opposed to textual duplication) even a low similarity score does not mean there is no plagiarism. We always bear in mind that an anti-plagiarism tool cannot detect all problems; the additional work carried out in the checking process, and the peer reviewers' comments on originality and innovation, are still the most important factors to ensure the originality of a paper. Figure 5.3 shows in graphic form the general rules applied in our own journals.

Case Study

In this case we asked the permission of the author and the reviewer to discuss this paper as an example of how to write the methods section without too much repetition.

A manuscript submitted to *Journal of Zhejiang University-SCIENCE B (Biomedicine and Biotechnology)* had gone through the peer review process; however, before making the final decision on publication, it is the journal's practice to run a CrossCheck. In this instance, the check revealed verbatim repetition of the description of the methodology used from the author's own previous publications, with only partial citation. A comparison of the similar sections is shown in Fig. 5.4a.

In the paper in question, the author used the same methods she had described—to study a different topic—in her published papers. CrossCheck found more than 1000 words in the methods section which had been copied verbatim from four previously published papers by the same author. The similarity indices of these four papers were 15, 6, 3, and 1 %. The author was therefore asked to make revisions to avoid self-plagiarism.

In her first revision, the author completely replaced the description of the method with a citation, as shown in Fig. 5.4b. This avoids the problem of self-plagiarism, but the absence of any detail makes the method section uninformative to the reader. One of the reviewers was asked for advice; his response was as follows:

(a)

2.3 Myrosinase activity determination Myrosinase activity was determined as described previously by Yuan et al. (2009). Broccoli sprouts (0. 5 g) were homogenized with 1. 8 mL of 50 mM of MES buffer (pH 6.0) in an ice bath, incubated at room temperature for 5 min, and centrifuged at 10, 000 × g and 4 °C for 10 min. The supernatants were collected and used for measurements. The assays were conducted with 1 mM sinigrin and 20 μL of supernatants in a total volume of 100 μL. After incubation at 37 °C for 15 min, the reaction was stopped by boiling (100 °C for 5 min). The amount of glucose formed by myrosinase was measured using Glucose GOD/PAP Kit (Shanghai Rongsheng Biotech Inc., Shanghai, China). The myrosinase activity was expressed as nmol glucose formed per minute and mg total protein.

(b)

2.2 Glucosinolate assay

Glucosinolates were extracted and analyzed as previously described (Yuan *et al.*, 2009).

2.3 Myrosinase activity determination

Myrosinase activity was determined as described previously by Yuan *et al.* (2009).

2.4 Sulforaphane measurement

Sulforaphane content was determined as described previously by Guo, Yuan & Wang(2011).

(c)

2.2 Glucosinolate assay

Glucosinolates were extracted and analyzed as previously described with minor modifications (Yuan *et al.*, 2009). 500 mg sprouts were boiled in 3 mL water for two times respectively, and 1 mL of the combined aqueous extract was applied to a DEAE-Sephadex A-25 (35 mg) column (GE Healthcare, Piscataway, NJ). The desulphoglucosinolates were obtained according to the procedure of Yuan *et al.* (2009). Then the extraction was analyzed by high performance liquid chromatography (HPLC). The glucosinolate concentration was expressed as μmol/g fresh weight of broccoli sprouts.

2.3 Myrosinase activity determination

Myrosinase activity was determined using spectrophotometry as described previously by Yuan *et al.* (2009). 500 mg broccoli sprouts were ground with 1.8 mL of 50 mM MES buffer (pH 6.0) to homogenate at 0°C, and then incubated at 25°C for 5 min. After centrifuging at 10,000 rpm and 4 °C for 10 min, the supernatants were collected for measurements. The assay were followed the procedure of Yuan *et al.* (2009). The myrosinase activity was expressed as nmol glucose formed per minute and mg total protein.

Fig. 5.4 A case study of how to write the methods section without too much repetition **a** CrossCheck report highlighting areas of similarity to the author's previous publications; **b** The first revision of the methods section, with citation only; **c** The second revision of the methods section based on the suggestion of the reviewer, with rewording (including more details) as well as citation

> … If the method is exactly the same that [as] the original one, I suppose that may be correct to express in this way. But, usually there are minor modifications from previous methods. Indeed, in the manuscript I reviewed, [the] authors said "Glucosinolates were extracted and analyzed as previously described with minor modifications (Yuan et al. 2009)". Besides, even if the method is the same that [as] a previous one, as a reader I think [it] is interesting to see some things in the paper. As [an] example, [does she] use the extraction method resin column or not? If yes, I am not interested in the method, if not then I can read the original method. Are glucosinolates by HPLC or MS, or GC? Depending on the method I may be interested in the original paper. So, I think that could be correct, but I don't like it and I think [it] is better if more information appears in Mat & Meth. I hope that this thought can help you to take a decision.

Accordingly, the author was asked to make a second revision, providing proper citations and a clear but concise statement of the materials and methods used in each step, without too much surplus; this was satisfactorily done. Figure 5.4c shows part of the method section. From such communication, the author showed that she would pay more attention to the method section to avoid repetition from now on.

From this case study, maybe authors will give enough information to avoid being accused of duplication in this section.

Practical Lessons for Authors and Editors

The first study reprinted above [1] included a survey of the practice of leading journals. They demonstrated a range of different ways of avoiding the verbatim repetition of previously published descriptions of methods. When the method is identical to that described in a previously published article (whether by the same or different authors), it is preferable to cite the previous article; if the wording is repeated verbatim, it must always be credited to the original publication. Alternatively, a phrase such as 'by standard method' or 'as previously described' may be used, as in *The Lancet* etc. [19]. However, if the precise description of the method is especially important for understanding the article, the attachment of supplemental material may be preferable, as in *Science* (Fig. 5.5). Only if the methods used are totally novel is it necessary to include a detailed description.

One of *JZUS B's* reviewers, Dr Jonathan McCanless, from the USA, also provided the following more detailed suggestions for biomedicine and biotechnology authors:

1. 'If using kits and supplier is referenced, then this is enough, due to most suppliers providing manuals.
2. If using traditional/homemade methods, include [the] following statement: "based on methods previously described" and include reference(s), as these procedures have all been documented in early chemistry/biochemistry journals … this is how we know how to do them.
3. If the procedure is novel, or incorporates a novel step to a known method, provide details. (If the entire procedure is novel, then a separate publication

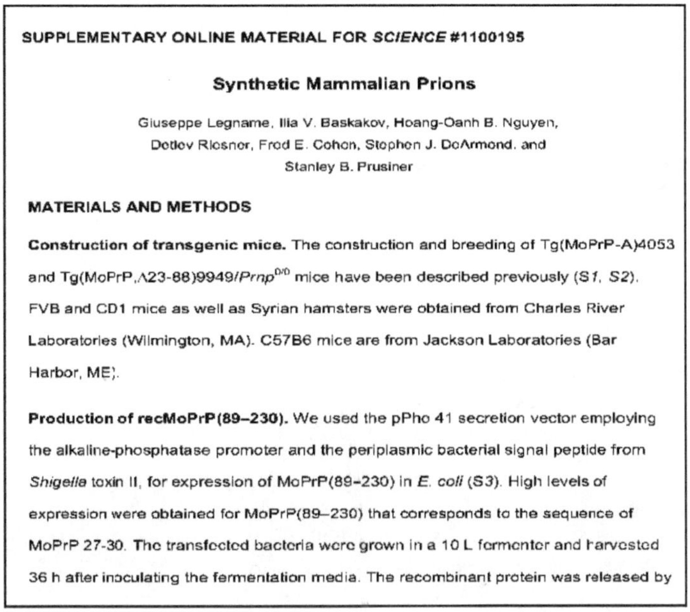

SUPPLEMENTARY ONLINE MATERIAL FOR *SCIENCE* #1100195

Synthetic Mammalian Prions

Giuseppe Legname, Ilia V. Baskakov, Hoang-Oanh B. Nguyen, Detlev Riesner, Fred E. Cohen, Stephen J. DeArmond, and Stanley B. Prusiner

MATERIALS AND METHODS

Construction of transgenic mice. The construction and breeding of Tg(MoPrP-A)4053 and Tg(MoPrP,Δ23-88)9949/*Prnp*$^{0/0}$ mice have been described previously (*S1*, *S2*). FVB and CD1 mice as well as Syrian hamsters were obtained from Charles River Laboratories (Wilmington, MA). C57B6 mice are from Jackson Laboratories (Bar Harbor, ME).

Production of recMoPrP(89–230). We used the pPho 41 secretion vector employing the alkaline-phosphatase promoter and the periplasmic bacterial signal peptide from *Shigella* toxin II, for expression of MoPrP(89–230) in *E. coli* (S3). High levels of expression were obtained for MoPrP(89–230) that corresponds to the sequence of MoPrP 27-30. The transfected bacteria were grown in a 10 L fermentor and harvested 36 h after inoculating the fermentation media. The recombinant protein was released by

Fig. 5.5 Attachment sample from *Science* http://www.sciencemag.org/content/305/5684/673

based on the method would be justified; if minor alterations are performed then state "based [on] previously detailed protocols with the exception of …" Provide a reference for basic protocol, and detail [any] innovative changes that were required to complete your experiment.)'

In *JZUS's* guidance on ethics issues (given in full in Chap. 10), we define 'Replication of methods sections in Biomedical journals without clear statement of the source' as one of nine basic forms of plagiarism as shown below.

Replication of Methods Sections in Biomedical Journals Without Clear Statement of the Source: When a standard method is identical to that described in a previously published article (whether by the same or different authors), it may be repeated verbatim, but a full citation must be provided to the original source; otherwise, the behavior will be considered plagiarism.

Remedy—identify the quoted method and provide full citation to the original source. If the description is particularly long or complex, it may either be (a) appended to the paper as supplemental material or (b) provided in the form of a link to the published original, if this is freely available online.

References

1. Jia, XY., XF. Tan, and YH. Zhang. 2014. Replication of the methods section in biosciences papers: Is it plagiarism?. *Scientometrics* 98: 337–345 (with permission of Springer).
2. Zhang, YH., XY. Jia, HF. Lin, and XF. Tan. 2013 Editorial: be careful! avoiding duplication: A case study. *Journal of Zhejiang University-SCIENCE B (Biomedicine & Biotechnology)* 14 (4): 355–358. (with permission of JZUS).
3. Zhang, YH., and XY. Jia. 2013. Republication of conference papers in journals?. *Learned Publishing* 26(3): 189–196(with permission of Springer).
4. Council of Science Editors (CSE). 2012. CSE's white paper on promoting integrity in scientific journal publications. 3.0 Identification of research misconduct and guidelines for action. http://www.councilscienceeditors.org/i4a/pages/index.cfm?pageid=3643#313 (accessed Sept 18, 2012).
5. Roig, M. 2012. Teach scientist to paraphrase. *Nature* 481: 23.
6. Roig, M. 2009. Plagiarism: Consider the context. *Science* 325: 813–814.
7. Committee on Publication Ethics (COPE). 2012. http://www.publicationethics.org (accessed September 20, 2012).
8. Zhang, Y.H. 2010. CrossCheck: An effective tool for detecting plagiarism. *Learned Publishing* 23: 9–14.
9. Zhang, Y.H., and X.Y. Jia. 2012. A survey on the use of CrossCheck for detecting plagiarism in journal articles. *Learned Publishing* 25: 292–307.
10. Journal of Zhejiang University SCIENCE. 2012. International reviewers. http://www.zju.edu.cn/jzus/reviewer.php#B (accessed September 2, 2012).
11. Zhang, Y.H. 2010. International, not 'campus', please. *Nature* 467: 789.
12. Thomson Reuters. 2012. https://jcr.incites.thomsonreuters.com.
13. Miller, N.R. 2011. Checking for plagiarism, duplicate publication, and text recycling. *Lancet* 377: 1403.
14. Burnette, W.N. 1981. "Western blotting": Electrophoretic transfer of proteins from sodium dodecyl sulfate-polyacrylamide gels to unmodified nitrocellulose and radiographic detection with antibody and radioiodinated protein A. *Analytical Biochemistry* 112: 195–203.
15. Meddings K. 2012. CrossCheck plagiarism screening: What's the magic number? http://www.ismte.org/File/Aug11CrossCheck.pdf (accessed June 3, 2012).
16. Kallet, R.H. 2004. How to write the methods section of a research paper? *Respiratory Care* 49: 1229–1232.
17. Wager E. 2012. How should editors respond to plagiarism? COPE discussion paper. http://publicationethics.org/files/COPE_plagiarism_discussion_%20doc_26%20Apr%2011.pdf (accessed on August 18, 2012).
18. Wager, E., S. Fiack, C. Graf, A. Robinson, and I. Rowlands. 2009. Science journal editors' views on publication ethics: Results of an international survey. *Journal of Medical Ethics* 35: 348.
19. Sarwar, N., AS. Butterworth, and DF. Freitag et al., 2012. Interleukin-6 receptor pathways in coronary heart disease: A collaborative meta-analysis of 82 studies. *The Lancet* 379(9822): 1205–1213. http://www.thelancet.com/journals/lancet/article/PIIS0140-6736(11)61931-4/fulltext.

Chapter 6
Computing and Electrical and Electronic Engineering: Republication of Conference Papers

The republication of conference proceedings as journal articles seems to be particularly prevalent in the areas of Computing and Electrical and Electronic Engineering. Our own use of CrossCheck turned up a disproportionate amount of duplication of this type in these particular disciplines; we also noted in our previous survey (see Chap. 2) that the attitudes to this question of authors in this area were markedly different from those in other fields. We therefore felt that it would be useful to carry out a survey of the attitudes of the editors of relevant journals to this issue; we surveyed over 300 journals in the field, and the results were published in an article entitled 'Republication of conference papers in journals?', which is reproduced in full (with the publisher's permission) below [1].

Introduction

It is well known that conference proceedings play a much larger role in publishing and communication in both Computer and Electrical and Electronics Engineering (EEE) sciences than in other fields [2–7]. However, it is unclear to what extent journal editors in these areas accept articles for publication that have been previously published as conference papers. If editors rely on CrossCheck [8, 9] to detect possible plagiarism, they are likely to find papers with a high similarity score simply because they have been previously published as conference papers; technically, this would be defined as self-plagiarism.

In 2011, the present we carried out a global survey of authors in a range of disciplines [8]. One of the 22 questions was: 'Should papers previously published in conference proceedings legitimately be republished in journals?' 60 % of the respondents, across a range of different disciplines, thought that conference papers could properly be republished provided that the author included new content; on average, they indicated that new material should constitute 46 % of the revised paper. However, 22 % of the respondents considered this to be duplicate publication, even with the addition of new content.

Nevertheless, there were clear subject differences. In the field of Computer Science/Electrical Engineering in particular, only 1 respondent (5 %) indicated that

Y.H. Zhang, *Against Plagiarism*, Qualitative and Quantitative Analysis of Scientific and Scholarly Communication,
DOI 10.1007/978-3-319-24160-9_6

such papers should be rejected out of hand (the lowest score in any field), while 17 (85 %) indicated that it was acceptable to republish a proceedings paper provided that it included new content. However, since the sample was too small to be statistically significant (only 20 respondents to the survey were from the field of Computer Science/Electrical Engineering), it was felt necessary to carry out a more in-depth investigation of the issue. Over 300 journals in the field were therefore surveyed about their policy.

Methods

Journals were identified by searching seven Computer Science categories (Artificial Intelligence; Cybernetics; Hardware and Architecture; Information Systems; Interdisciplinary Applications; Software Engineering; and Theory and Methods) in Thomson Reuter's Journal Citation Reports (JCR) that was used as the source since inclusion in its database is an indication of journal prestige, and its journals were selected from the leading publishers and societies in the field. A total of 615 journals were listed in these categories. From these and based on the publishers' website conditional access for the editorial information, 323 journals were selected. They included 123 journals from the Institute of Electrical and Electronics Engineers (IEEE), and 96 published by Elsevier. A total of 963 editors-in-chief, handling and/or managing editors were sent an email between 19 July–2 August 2012, inviting them to participate in the survey on SurveyMonkey (http://www.surveymonkey.com).

The survey contained four questions:

Q1. Do you accept for publication articles which have previously been published in conference proceedings? (Yes/No)
Q2. If YES to Q1, do you require the author(s) to make changes to the paper before republication? (Yes/No)
Q3. If YES to Q2, please outline the nature of the changes. (Free text response) Why do you consider that these changes are necessary? (Free text response)
Q4. If NO to Q1, why not? (Free text response).

Results

Respondents

120 of the 963 editors (12.5 %) of 78 of the 323 journals (24.1 %) answered one or more questions. As Fig. 6.1 shows, the majority of respondents are from Western counties, with only 9 from Asia.

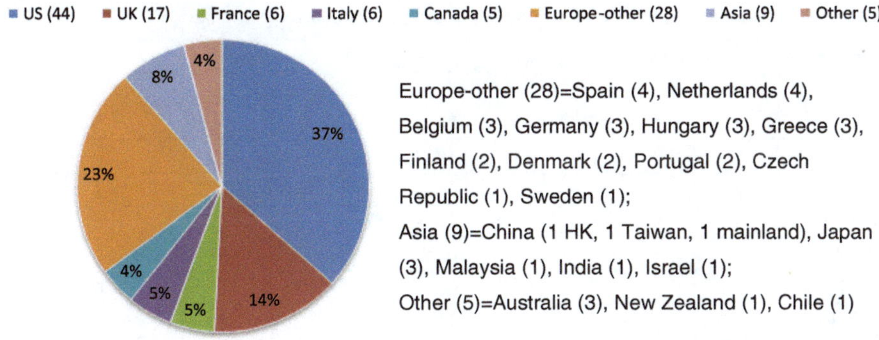

Fig. 6.1 Geographic distribution of respondents to the survey (n = 120)

Cross-analysis of journal and publisher by Q1's respondents showed that journals are from 10 different publishers: Elsevier, IEEE, Springer, Association for Computing Machinery (ACM), Massachusetts Institute of Technology (MIT), World Scientific, Cambridge University Press (CUP) Wiley, Morgan Kaufmann and Zarqa University. The largest number of both journals and editors was from Elsevier (79 editors of 39 journals), followed by IEEE (17 editors of 16 journals) (see Table 6.1).

Responses

Q1. Do you accept for publication articles which have previously been published in conference proceedings? (Yes/No).

For Q1, 120 editors of total 78 journals responded to this question as shown in Table 6.1. 93 (77.5 %) editors of 53 journals chose 'Yes' and 27 (22.5 %) editors of 25 journals chose 'No'. There were interesting differences between the two publishers with the most journals represented in the responses to this question: 18 (22.8 %) of the 79 editors of 17 (43.6 %) of the 39 Elsevier journals, but only 4 (23.5 %) of the 17 editors of 4 (25 %) of the 16 IEEE journals chose the answer 'No'.

{Note that 7 of the 27 respondents to Q4 (who had answered 'No' in Q1) clearly indicated that they would, in fact, publish conference papers which had been significantly extended with new material; if their responses would be construed as 'Yes', the percentage saying 'Yes' is actually 83.3 % [(93 + 7)/120], and the percentage saying 'No' is 16.7 % [(27 − 7)/120].}

Q2. If 'Yes' to Q1, do you require the author(s) to make changes to the paper before republication? (Yes/No).

93 editors of 53 journals responded to this question. 92(99 %) answered 'Yes' and just one answered 'No'.

Table 6.1 Responses to Q1 by editor and by journal, grouped by publisher

Publisher	Answer to Q1					
	Editor			Journal		
	Total	Yes	No	Total	Yes	No
Elsevier	79	61	18	39	22	17
IEEE	17	13	4	16	12	4
Springer	8	6	2	8	6	2
ACM	5	5	0	5	5	0
World Scientific	3	3	0	3	3	0
MIT Press	3	2	1	3	2	1
CUP	2	0	2	1	0	1
Wiley	1	1	0	1	1	0
Morgan Kaufmann	1	1	0	1	1	0
Zarqa University	1	1	0	1	1	0
Total, *n* (%)	120	93 (77.5 %)	27 (22.5 %)	78	53 (68 %)	25 (32 %)

Q3 If 'Yes' to Q2, please outline the nature of the changes; why do you consider that these changes are necessary? (2 parts; free text responses).

91 editors of 53 journals responded to this question. One made clear that the aim of his publication is to abbreviate papers of general interest for its specific (magazine) readership—this response has been considered a negative response in the following analysis. One other stated that, while the journal would not republish papers previously published in conference proceedings, it would publish expanded versions of posters, short papers etc. from such proceedings, while another stated that only conference papers previously published by the same publisher (a society) would be considered. When studying the 91 free text responses in detail, the authors observed a number of particularly frequently-occurring words and phrases. Frequency analysis was therefore carried out for a number of these words and phrases. For example, in the responses to the first part (Pt1) of the question, 'new' occurred 43 times; '%' occurred 40 times; 'more' occurred 27 times; 'exten*' (extend, extension etc.) occurred 19 times; 'add*' (adding, added, additional etc.) occurred 18 times; 'result/results' occurred 17 times; 'substanti*' (substantial, substantive) occurred 16 times; 'detail' occurred 12 times; and 'experiment' occurred 10 times.

In the responses to the second part (Pt2) of the question (where respondents where asked to give their reasons for the changes requested), the need to avoid duplicate publication was mentioned 43 times. The more rigorous peer review requirements of journals were mentioned 11 times; the length constraints of proceedings were mentioned 8 times; and copyright issues were mentioned 7 times.

Table 6.2 indicates the number of times each of the key words or phrases occurred in responses to Q3. Further examination of the responses in which these key words and phrases occurred made it possible to classify them as follows:

Table 6.2 Occurrence of key words and phrases in responses to Q3 with two parts

Question 3	Please outline the nature of the changes	Why do you consider that the changes are necessary?
Term	Occurrences in Pt 1	Occurrences in Pt 2
New	43	4
Same + duplicat* + double + twice + republi* + re-publi*	4 (1 + 0 + 0 + 0 + 2 + 1)	43 (16 + 7 + 2 + 4 + 10 + 4)
%	40	0
More—more than	25 (27 − 2)	18 (18 − 0)
Exten*	19	3
Add*	18	10
Result*	17	8
Substanti*	16	5
Detail*	12	8
Review + referee	8 (7 + 1)	11 (11 + 0)
Experiment*	10	1
Proof*	9	1
Significant	9	0
Complete	7	9
Value	7	9
Change*	8	8
Theor*	8	1
Long + length + page	5 (2 + 1 + 2)	8 (2 + 0 + 6)
Quality	2	8
Expan*	7	2
Improve	7	2
Copyright + legal	5 (5 + 0)	7 (6 + 1)

(continued)

Table 6.2 (continued)

Question 3	Please outline the nature of the changes	Why do you consider that the changes are necessary?
Differ*	6	6
Expla*	6	2
Compar*	4	0
Original	4	1
Depth + deep	4 (3 + 1)	1 (0 + 1)
Related	3	1
Evaluat*	3	1
Technical	3	0
Related	3	1
Example	3	0
Thorough	2	3
Algorithm	2	0
Feedback	2	0
Discuss*	2	1
Self-plagiarism	1	0
Title + name	2 (1 + 1)	1 (1 + 0)

*: Prefix search

Q3 Part 1—Changes Required

Inclusion of new content
81 (90 %) of the 90 positive respondents made explicit mention of the necessity to include new or extended content. 38 respondents mentioned a specific minimum percentage of new content. The most commonly mentioned percentage (22 respondents) was 30 %. One mentioned 75 %, two 70 %, three 60 %, five 50 %, two 40 %, one 35 % and two 20 %. (It should be noted in addition that 7 respondents who had said 'No' to Q1 actually indicated, in their responses to Q4, that they would publish a significantly expanded version of a conference paper; of these, one mentioned 40 % and one 50 %.)

While some stressed the need for more detailed discussion, background, etc., others specifically mentioned the need for additional results, proofs, etc. In particular, 18 (20 %) particularly mentioned the need for additional or expanded experimental or theoretical results, and 9 (10 %) the need for more or extended proofs.

It was mentioned repeatedly that conference papers are generally little more than an 'extended abstract', often prepared in haste and arguably with lower standards of acceptance: 'A conference paper (with page limit) considered as an extended abstract should be published in a Journal in a complete form.'

As one respondent made clear, '[the] original paper needs to be cited and differences explained'. One respondent also mentioned the importance of using original text, rather than merely cutting and pasting; another specifically mentioned avoidance of self-plagiarism.

Q3 Part 2—Reasons for Changes

83 respondents meaningfully addressed the second part of the question. Their responses could be classified under six main headings:

1. *Adding value/improving quality*

43 responses (51.8 %) mentioned that the journal version should be of higher quality and thus of greater value to the community: 'Extended paper should add more value, more information, and more details'; 'the journal's role is to publish expanded, more thoroughly-developed versions of the research'.

2. *Avoiding duplicate publication*

The avoidance of duplicate publication, since it served no useful purpose for readers, was explicitly mentioned by 28 respondents (33.7 %): 'Re-publishing the same paper would be duplicating content that is already available'; 'the conference version is already peer reviewed and archived'; 'No point [in] having the same article twice because conference proceedings are now available online'; 'Conference papers are easily accessible on the web. No need to republish.'

3. *Overcoming the constraints of proceedings*

14 respondents (16.9 %) explicitly mentioned the constraints on extent, style etc. imposed by proceedings publication. They saw publication of the full paper in the journal as the way to overcome these constraints: 'Conference papers have obviously limited number of pages and quality of journal papers should be better than conference ones'; 'Conference publications tend to be short and therefore incomplete. They also tend to be preliminary in nature, sometimes errors are present'; 'The aim of 'short' submissions is different from that of journal articles but often is up-to-the-minute work which can be put in context and expanded upon in a journal submission.'

4. *Higher peer-review standards*

6 respondents (7.23 %) mentioned that the paper would go through a more rigorous peer review process: 'reviews are rarely as thorough as for a journal'; 'The conference review process is hurried and light, concentrating on innovation rather than long-lasting significance or importance'; 'Permanent archival journals require rigorous attention to details to satisfy expert reviewers'; 'the standards for journal publication are higher.'

5. *Copyright issues*

Copyright issues were mentioned by 5 respondents (6 %): 'conference papers in CS are proper publications—copyright resides with the publisher in many cases'.

6. *Journal policy*

3 respondents (3.6 %) stated that the required changes were the journal's or society's official policy

(See the full text of 91 responses is included in Appendix 1).

Q4. If 'No' to Q1, Why not? (Free Text Response)

27 editors from 25 journals answered this question. Many of the respondents appear to have understood Q1 slightly differently from those who answered 'Yes' to Q1, assuming that it referred to verbatim republication. Perhaps unsurprisingly, therefore, their responses were frequently the obverse of the responses of those who answered 'Yes' to Q1; in 7 cases (26 %) the publication policies one can deduce from their responses are to all intents and purposes identical (i.e. they would not republish verbatim, but they would publish a significantly different and expanded version)—see point 2 below.

In their responses, terms referring to duplicate publication were mentioned 16 times (Table 6.3). The journal's insistence on publishing only novel/original material was mentioned 6 times. The policy of the publisher of the journal and/or of the original conference proceedings was mentioned 4 times. Copyright issues were mentioned twice. The percentage of new material required to justify publication was also mentioned twice: one respondent specified more than 50 %, the other

Table 6.3 Occurrence of key words and phrases in response to Q4

Term (key words)	Occurrences
	Q4: If 'No' to Q1, why not?
Already + twice + duplicat* + republi* + recycling	16 (5 + 2 + 3 + 5 + 1)
Novel + original	6 (1 + 5)
Polic*	4
Copyright	2
%	2
Self-plagiarism	1

*: Prefix search

40 %. One respondent specifically mentioned the issue of self-plagiarism (see the full text of 27 responses is included in Appendix 2).

Their reasons fell into the following categories (points):

1. *Duplicate publication (mentioned in 16 responses)*

'Since conference proceedings are already widely available on the Internet'; 'A paper should not be published twice in different media.' Note that duplicate publication was also mentioned in 28 responses to Q3-Pt2 (those who replied 'Yes' to Q1). Thus 28 + 16 = 44 respondents (36.7 %) in total mentioned the necessity of avoiding duplicate publication.

2. *Would publish, but only if expanded with new material (7 responses)*

'We only accept such papers if it can be demonstrated that more than 50 % of the material is new.' Note that the same stipulation was found in 81 responses to Q3-Pt2 (those who replied 'Yes' to Q1). Thus in total 81 + 7 = 88 (73.3 %) of respondents mentioned the need to include new material.

3. *Prefer original papers (7 responses)*

'We prefer original conceptual papers which are innovative'; 'we publish only original contributions.'

4. *Journal policy (3 responses)*

'We follow the IEEE policy on self-plagiarism'.

5. *Copyright issues (3 responses)*

'Often conference proceedings require transfer of copyright, and thus we cannot publish the same material'. Clearly these respondents have understood the question to refer to verbatim republication.

6. *Only publish conference papers in specific circumstances (2 responses)*

'These would be part of special issues based on the best papers in a specific conference'; 'We would typically set up special journal issues, peer-review the papers, and published the ones that were acceptable.'

The full text of all 27 responses from 25 journals is included in Appendix 2.

Discussion

As noted, in the field of computer science, conference proceedings are known to be one of the most important means of communicating one's findings, although in current years, there has been some debate on this [10–15]. Vardi, the editor-in-chief of the *Communications of the ACM,* wondered 'whether we are driving on the wrong side of the publication road' [13]. The republication of substantially the same papers in journals used to be common practice, but is now increasingly discouraged [16, 17]. Duplicate publication and self-plagiarism (even when the original publication is properly cited) are considered poor academic publishing [8, 16, 18]; the tools to detect them are now available with Crosscheck [7, 19]. As one of the respondents to this survey noted: 'before the mid-1990s, many societies encouraged their members to present an earlier version of their papers at the society conferences (and thus to publish them in the conference proceedings) before they submitted the full paper to the society transactions. Examples include the Institute of Electrical and Electronics Engineers (IEEE) Control Systems Society etc., with which I am involved. If you check out their publications in the 1970s–1990s, you will find that many papers acknowledged earlier publication of the same papers in International Federation of Automatic Control (IFAC) conference proceedings. Since the mid-1990s, however, the situation has gradually changed. Many societies (for example, IEEE as a whole, including the above-mentioned societies) started to discourage their members from doing this. However, due to long-established habits, many authors—especially those from older generations—are still doing the same, or something similar. Likewise, most societies are not very strict on limiting this common practice.'

Indeed, in early 2000, COPE met a typical case on duplicate publication based on conference proceedings for journal editor reference [17].

In current years, many journals have had their explicit policies on this point [16, 18] for example, *Artificial Intelligence Journal* on its Website stresses "In particular, a previous conference publication by the same authors does not disqualify a submission on the grounds of novelty…". And as a leading publisher, Elsevier also has its strict policy on conference paper's resubmission for its journal editors to carry on [20].

Analysis of the responses to this survey indicate very clearly that journal editors are almost unanimously opposed to the republication of conference papers in exactly the same form: 'there is no need to publish the same paper twice, the journal provides a longer format for more complete presentation; that is its purpose.' Their reasons are mainly to do with avoiding duplicate publication, self-plagiarism and copyright problems. As Table 6.1 shows, 32 % of responding journals absolutely decline to republish conference papers under any circumstances. The other 68 % are willing to publish a version of papers that have previously appeared in conference proceedings, but 99 % of these insist that the authors must change the content, adding valuable new material. They generally insist on a significant amount of new material (actual percentages mentioned range from 20 to 75 %—most commonly 30 %).

Conclusions

Republication of a paper which has previously appeared in conference proceedings has long been relatively common and widely accepted in the field of computing and electrical and electronics engineering. However, some journals now question the value of doing so and prefer to focus their content on completely novel work. Those that do republish conference papers will generally only do so if the paper has been substantially reworked to include additional detail which could not be included in the conference paper, such as detailed proofs or wider comparison with other work in the field. The reasons put forward are that duplication of publication is unnecessary, and that journals have higher standards, and more space, in order to provide their readers with greater value.

However, the majority of those respondents who named a figure felt that as much as 70 % of the paper could remain unchanged from the previously published version, which many would consider self-plagiarism. It is thus inevitable that plagiarism detection tools such as CrossCheck will come up with very high similarity scores for such papers. It follows that editors in this field, in particular, cannot rely on CrossCheck, but must use their reviewers' and their own judgment to determine whether or not a version of paper previously published in conference proceedings has sufficient additional value to warrant publication (it goes without saying that the original publication must be properly cited in all cases).

Practical Lessons for Authors and Editors

As our survey results show, there is little uniformity in policy between journals in the fields of Computing and EEE. We believe that consistent policies are long overdue, and our own journal guidelines [21] follow (more detail will be found in Chap. 10).

Republication of Conference Papers with Little Added Value: If the paper has not been substantially reworked to include additional detail which could not be included in the conference paper, such as detailed proofs or wider comparison with other work in the field, it would be considered self-plagiarism.

Remedy—add 60 % or more of substantive new material, which adds value to the original conference paper. Full citation to the original publication must be given, as well as copyright permission from the original publisher. [1]

Appendices: Free Text Responses to Question 3 and Question 4

Appendix 1: Free Text Responses to Question 3

No.	Q3-Pt1: Please outline the nature of the changes (n = 91)	Q3-Pt2: Why do you consider that these changes are necessary? (n = 87)
1	Should be 70 % changes to the paper published in conference proceedings. The changes should include detail out related work, further evaluation of method	No point of having the same article. The most important thing is the dissemination of knowledge
2	At least 30 % across most aspects	The journal paper should be an improvement of the conference one
3	Expanding out all omitted materials in the Proceedings due to space constraints	The journal paper should be complete
4	30 % new material	Conference proceedings are now available on line
5	The overlap shouldn't be more than 50 %	The journal version of a conference paper should include much more details and insights
6	At least 30 % new materials added	Why publish the same paper in the different forms?
7	Substantial additions, more thorough coverage, results, comparisons, etc.	So that the reader will benefit from reading the paper (having read the conference proceedings)
8	The paper must contain contents largely expanded to cover the advancement in the topical subject or field of study, typically with 70 % percent more or different materials compared to the conference paper	Mandatory!
9	Significantly more experiments, explanation, and algorithmic details	There is no need to publish the same paper twice. The journal provides a longer format for more complete presentations; that is its purpose
10	Improve the quality of the papers, 30 % new	Conference papers are inevitably in complete and of lower quality than what are needed for a journal article
11	Add 30 % new content	Explain the work in more details, deepen the theory and report more experimental results
12	The journal submission must contain 50 % new materials	

(continued)

(continued)

No.	Q3-Pt1: Please outline the nature of the changes (n = 91)	Q3-Pt2: Why do you consider that these changes are necessary? (n = 87)
13	I believe that an Extension with more details, theory, examples, concepts, results, experiments and comparisons are required	
14	At least 20 % original material	To respect the copyright
15	The journal paper needs to be a mature, well-founded contribution that has been fully validated either theoretically or experimentally. Usually, the conference paper only outlines the contribution, thus it needs to be substantially Extended with proofs and/or Complete experimentation	To obtain a Complete self-standing contribution for the future, which other works can build on
16	There has to be at 30 % new material in the journal version	There should be added value otherwise the authors are simply publishing the same paper twice
17	To have more Complete version and to improve quality	For journal should have more complete and better quality papers
18	Detailed proofs and full explanation should be given	A conference paper (with page limit) considered as an Extended abstract should be published in a Journal in a Complete form. My area is mathematical, and journal review is mandatory to be recorded as a correct result in the history
19	Substantial extension, new experiments, etc.	Otherwise it is a publication duplicate, this should never be accepted!
20	Expand and update discussion	These changes make the paper better; conference proceedings have unnatural constraints
21	1. At least 25 % new material 2. Revisions requested by reviewers	(1) is the policy of the Association for Computing Machinery
22	Significant changes/enhancements (min 30 % enhancements, often more)—both conceptually and in details/evaluations. Publications must add substantial value adds	We really do not Republish the papers but would consider papers that have substantial new material that builds upon a prior conference/workshop paper
23	Extended research results, validation, evaluation	Substantiate more early research results
24	Submissions can contain material published in one or more conference papers. Submissions should contain at least 20 % percent by length of material that is unpublished. This new material should make substantial new contributions. This would typically be additional theoretical or experimental results, or both	Strengthen the quality and significance of the previously published material

(continued)

(continued)

No.	Q3-Pt1: Please outline the nature of the changes (n = 91)	Q3-Pt2: Why do you consider that these changes are necessary? (n = 87)
25	Major Extensions, and avoid self-plagiarism	A paper published in a conference is to be considered archival. Also, given widespread access to internet, it is easily available. Therefore, no reason to duplicate publications
26	We require significant Extension, such as new results and/or discussion, to provide substantial 30 % 'value added' to the journal version. We also require that the paper be revised, if necessary, to be sufficiently complete and thoroughly documented for Archival journal publication	The conference paper is already published and available. In addition, the standards for journal publication are higher
27	The title may be the same, but the journal submission must be much more detailed and complete than a conference presentation. We rely on the opinion of expert reviewers to determine if the submission is worthy of consideration	Permanent archival journals require rigorous attention to details to satisfy expert reviewers. Almost nothing is accepted without a major revision
28	The proceedings should be an extended abstract	Otherwise it would be basically as in the proceedings and then the answer to 1 would be NO
29	Extension in theory, algorithm, and/or empirical validation	
30	Complete proofs of results, improvements after conference feedbacks	To have a self-contained presentation of results
31	Full proofs of Theorems, more complete examples, any additional material that could not be included in the conference paper due to strict page limits	Republication is allowed only because conferences impose page limits
32	Substantial changes with at least 30 % new material	Conference publications are already peer reviewed publications. To republish substantially the same work would not be honest, and would devalue the original publication
33	Must be significantly different with the conference paper	Conference papers are simple and quick publication, journals are more formal and for wider distribution and reference
34	More depth or additional material	Conference reviews are patchy, page limits are tight
35	conf proc are preliminary reports, to be refined for peer review; if we have to quantify … 40 % reuse is a threshold;	Goal is to preserve conferences as a place to discuss research in process; reviews are rarely as thorough as for a journal; expected that authors needs to revise
36	Substantial expansion of new content	So the journal version represents added value to the reader

(continued)

(continued)

No.	Q3-Pt1: Please outline the nature of the changes (n = 91)	Q3-Pt2: Why do you consider that these changes are necessary? (n = 87)
37	Additional theoretical and empirical results, additional consideration of related work	Conference proceedings are Archival in nature in CS; the same paper should not be published twice, and the journal's role is to publish expanded, more thoroughly-developed versions of the research
38	Longer paper, more results, feedback from conference	To make the idea more Complete
39	Improvement in contents and changes in writing	Yes
40	The paper must be a significant extension of a conference paper	We do not allow duplicate publication
41	60 % must be new	We cannot publish the same paper
42	Substantial changes and extension of the conference version, 30 % new	Conference papers in CS are proper publications—copyright resides with the publisher in many cases
43	At least 30 % new material	Re-publication of the same material would be the same as publishing the same paper in two journals: unfair and unethical. From a practical standpoint, it would also make the journal superfluous
44	The journal publication gives authors a chance to add details, proofs, and comments, to Expand and to polish their presentations, and to fix typos and errors, inevitable for deep works presented by deadline to the conferences, where no corrigenda are available and the reviewers have too little time; the authors should use this chance; besides I should have some support to counter possible objections from some excessively zealous reviewers who cannot see the need for journal publication of proceedings papers and blindly apply the copyright law, not caring about the best interests of the field	The reasons are explained above
45	We require sufficient changes that someone finding both the journal article and the conference publication would receive value from the journal article over the conference publication. In other words, there has to be sufficient additional, new material to justify publication in the journal	To justify the second publication
46	Substantial amount (e.g. 30 %) of new material/work	Not to republish things already published

(continued)

(continued)

No.	Q3-Pt1: Please outline the nature of the changes (n = 91)	Q3-Pt2: Why do you consider that these changes are necessary? (n = 87)
47	Significant new and/or expanded material must be included	To avoid duplicative publication of essentially the same material
48	I'm answering this as if I were EIC of IEEE S&P Magazine, which I was until January 2011. This publication is a magazine, not a journal. Hence it aims to publish papers of general interest and not, in general, cutting edge research. It is subject to the IEEE's rules on republication. Usually, articles that are republished in the magazine require shortening and editing to make them suit the magazine's audience	The changes are needed to tailor the article's content to the magazine's audience
49	The article must deal the topic more in details, with new results	While journal articles have a wider diffusion, conferences (especially the main ones) a becoming more and more accessible via the web
50	Some technical details	Quality
51	Substantial Extension	Conference papers generally are short and do not have the breadth of a journal paper
52	Substantially new technical material, >60 %	Yes
53	30 % new material	Added value of the new publication
54	Give more explanations if needed	To overcome the page limit with conferences; and because often sufficient details are needed to make the work reproducible by readers
55	At least the paper should be 75 % different from the one published	Otherwise, this constitutes double submission
56	The paper is sent to referees for a more formal review by experts in the subject area	Typically, the conference proceeding version has been quickly reviewed by one editor, but did not go through the typical detailed review process
57	At least 30 % Substantive additional content	Otherwise no additional value is obtained from republishing the material
58	30 % extra material	Republishing only existing material gives no added value over the conference publication
59	Min. two new reviewers of the journal will check the improved conference paper again. If, for example, the paper was originally published at an ASME conference proceeding, then no copyright issues occur in case of re-publication in	Usually, the review processes of conference proceedings papers are not as strict as the journal reviews, also the page limits and the format requirements are different. Actually, someone may submit a conference paper unchanged, but to

(continued)

(continued)

No.	Q3-Pt1: Please outline the nature of the changes (n = 91)	Q3-Pt2: Why do you consider that these changes are necessary? (n = 87)
	an ASME journal. I do not accept a conference paper that was published by another publisher	have 2 papers with the same title and same content is strange—the authors would like to carry out some changes after having the experience at the conference
60	Add new material (at least 30 %, 1/3 of the paper must be new material)	Conference papers are typically shorter
61	It depends on the paper! More details of proofs, more experiment reports, development of the state of the art, etc.	Conference papers are too short to develop comprehensive presentation
62	30 % new material	
63	At least 30 % new material, significant results of lasting value	The material has already been published —same reason we do not republish articles from other journals. Also, copyright in at least some cases
64	Higher standard for acceptation	Higher quality
65	Substantial changes at least 40 % new content	Because the paper should be substantially different
66	Addition of significant new content	Otherwise a separate publication is not justified
67	Extension and depth	To improve the quality
68	There should be at least 50 % new material in the journal submission	To avoid copyright issues and also to emphasize that the motivation for conference papers and journal papers are different and their styles should be different as well as the amount of detail provided
69	30 % new material, either in the form of new theoretical material, or new experimental results	The journal version should be more extensive than the conference version
70	Extended results, e.g. more experiments, plus Extended related work, plus a more extensive description of the presented framework, or tool, or approach	The conference proceedings is already a referable publication per se, it makes no sense having it identically republished elsewhere
71	Provide full proofs of the results (which usually have been omitted from the conference version due to space limitations)	Journal publications are considered the definite publication of a result and must contain full proofs
72	Fuller account of data and methods, new analyses, and original text (i.e. not cut-and-paste)	Yes
73	In order to ensure copyright for both the conference and Elsevier we insist on a major rework—usually making the paper longer and more technical	Legal requirements and ensuring quality papers

(continued)

(continued)

No.	Q3-Pt1: Please outline the nature of the changes (n = 91)	Q3-Pt2: Why do you consider that these changes are necessary? (n = 87)
74	Paper should add more in depth information; paper should contain >50 % new material compared with the conference	We should avoid duplication (I mean copy and paste). Extended paper should add more value, more information, and more details. I do not accept cosmetic changes!
75	We will not publish full papers from conf. proceedings (usually ACM, or Springer LNCS) but will accept Substantially modified and extended submissions of Work-In-Progress, poster, short papers (<4 pages) published as part of such proceedings	The aim of 'short' submissions is different from that of journal articles but often contains up-to-the-minute work which can be put in context and expanded upon in a journal submission. Does not really count as a 'citable' publication as is often in an Appendix or 2nd volume of related conference publications
76	Typically require more experiments, proofs, details of implementation, additional to previous work	Conference publications tend to be short and therefore in complete. They also tend to be preliminary in nature, sometimes errors are present. The conference review process is hurried and light, concentrating on innovation rather than long-lasting significance or importance. Change for the sake of change is not necessary; thorough presentation of the material is what typically creates the need for changes to the conference paper in order to make it suitable for journal publication
77	New results must be included	To produce new results for researchers
78	I expect conference proceedings papers to be short versions of what would be more complete papers	I would not simply republish a conference paper, the journal paper would have to be more Substantive and contain a more Complete description of the work and additional results
79	More precise explanations, theoretical background, case studies, language improvements, etc.	Conference papers have obviously limited number of pages and quality of journal papers should be better than conference ones
80	Significant improvement or extension	Duplication adds no value
81	Extensions in contribution with clear explanation of the novelty of the 'new' paper, 50 %	We should not support re-publication of research; we should support publication of new and novel research contributions. There may also be copyright issues if re-publishing as well as problems with which version to cite. As an editor, I want other researchers to cite the paper in my journal, and I believe this is best achieved if a paper has an added value in relation to a previous publication

(continued)

(continued)

No.	Q3-Pt1: Please outline the nature of the changes (n = 91)	Q3-Pt2: Why do you consider that these changes are necessary? (n = 87)
82	Should be at least 30 % new material	There is no point having the same paper twice
83	The submission should contain a substantial amount of new results	To make the paper strong enough for journal publication
84	Papers have to be considerably expanded	for solving copyright issues
85	At least 30 % new material (RESULTS), different name, original paper needs to be cited and differences explained	Re-publishing the same paper would be duplicating content that is already available in ACM or IEEE digital libraries
86	Extend the paper	Journal paper should be more consistent
87	30 % more material	The conference version is already peer reviewed and archived
88	A different perspective and angle would generally be required or more detail	There is no point having an identical copy of the paper everywhere; problem is papers as conf articles are not cited or read anywhere near level of journals so journals are mandatory. Conferences are clearly important for interactive dialog
89	Provide full proofs of claims	Because journal publications should contain all details necessary to understand claimed results
90	30 % new material	It's common practice
91	More than 40 % new results and outcome (qualitatively)	My journal policy is set to not allow to have double archives documents, and also it is not appropriate to republish similar things

Appendix 2: Free-Text Responses to Question 4 (n = 27 Journals Editors)

No.	If NO to Q1, why not?
1	Because it has already been published. We only accept such papers if it can be demonstrated that more than 50 % of the material is new
2	A paper should not be published twice, in different media. Archival publication in a journal is a different topic than a conference paper. Of course, the journal article can speak about the same work presented in the conference, but the journal paper is of a very different nature and should never be a republication

(continued)

(continued)

No.	If NO to Q1, why not?
3	Duplicate publication
4	We publish only original contributions
5	These would be part of special issues based on the best papers in a specific conference. We do not accept individual papers which have been published in general conferences
6	Conference papers are easily accessible on the web. No need to republish
7	Our policy is here: http://www.ida.liu.se/ext/aijd/authors.html—AIJ considers eligible for publication mature high quality work that is Complete and novel. The question of whether a paper is complete is ultimately determined by reviewers and editors on a case-by-case basis Generally, a paper should include all relevant proofs and/or experimental data, a thorough discussion of connections with the existing literature, and a convincing discussion of the motivations and implications of the presented work. A paper is novel if the results it describes were not previously published by other authors, and were not previously published by the same authors in any archival journal. In particular, a previous conference publication by the same authors does not disqualify a submission on the grounds of novelty. However, it is rarely the case that conference papers satisfy the completeness criterion without the addition of new material. Indeed, even prize winning papers from major conferences often undergo major revision following referee comments before being accepted to AIJ. Authors should ensure that they have appropriate copyright permissions before submitting any material that has been previously published
8*	**Editor 1**. It is no longer original. It would only be accepted if a significant amount of additional material were added, such as reviewing other work, and a full evaluation. SO it would be ok if the conf paper formed the basis of a journal paper, but it would never be published 'as is' if the conference already had published it
9*	**Editor 2**. They have already published (*No.8 and No.9 in same journal)
10	Since conference proceedings are already widely available on the internet, there is no added value to having the same content appear in a journal as well
11	We follow the IEEE policy on self-plagiarism
12	Often conference proceedings require transfer of copyright, and thus we cannot publish the same material
13	In my field, conference proceedings are often considered archival, so publishing a conference paper again in a journal would be duplicate publication. We do allow articles that consolidate and extend work published previously in conference proceedings, however
14	Published papers should be updated, following suggestions from the audience
15	We prefer original conceptual papers which are innovative
16	Because they are already available. We would typically set up special journal issues, peer-review the papers, and published the ones that were acceptable
17	It is not necessary
18	We only consider full length original research articles
19	We publish 'original research'
20	Duplication
21*	**Editor 1**: It's a silly question. Common practice is to publish a paper RELATED to a conference paper, but having sufficient changes to be considered a new paper. (Otherwise the conference publisher would own the copyright and we wouldn't be able

(continued)

(continued)

No.	If NO to Q1, why not?
	to republish any way!) Typically, papers need 40 % or more of new material. They also need to be more rigorous than conference papers in various other ways, e.g., we'd normally expect all formal claims to be proved and all empirical claims to have evidence provided for them
22*	**Editor 2**: Publication is warranted only once—republishing material serves no legitimate purpose, and only inflates a person's list of publications with misleading information. (*No. 21 and No. 22 in same journal)
23	The journal is dedicated to publication of new results—not recycling of old results with minor updates
24	Too many authors then count the paper twice. It's a con. There is the occasional exception where the paper gets radically revised and becomes in effect a new paper, but most conference papers are not like this
25	It is necessary for authors to submit new papers that are derivative of their prior conference papers. Publishers (ACM vs. IEEE vs. Wiley) have policies and concerns, so we try to not cross them
26	The only reason for a journal publication is to provide a stamp of being reviewed. Conference papers are usually not thoroughly reviewed, in particular in terms of correctness
27	My journal policy is set to not allow to have double archives documents, and also it is not appropriate to republish similar things

References

1. Zhang, YH., and XY. Jia. 2013. Republication of conference papers in journals? *Learned Publishing* 26(3): 189–196. (with permission of Learned Publishing).
2. Cognitive Science Society. 2012. Policy on Journal Publication of Conference Papers. http://cognitivesciencesociety.org/conference_Archival.html#top.
3. Drott, M.C. 1995. Reexamining the role of conference papers in scholarly communication. *Journal of the American Society for Information Science* 46(4): 299–305.
4. Fortnow, L. 2009. Time for computer science to grow up. *Communications of the ACM* 52(8): 33.
5. Franceschet, M. 2010. The role of conference publications in CS. *Communications of the ACM* 53(12): 129–132.
6. Montesi, M., and J.M. Owen. 2008. From conference to journal publication: How conference papers in software engineering are extended for publication in journals. *Journal of the American Society for Information Science and Technology* 59(5): 816–829.
7. Patterson, D., L. Snyder, and J. Ullman. 1999. Best practice memo: Evaluating computer scientists and engineers for promotion and tenure; http://www.cra.org/uploads/documents/resources/bpmemos/tenure_review.pdf.
8. Zhang, Y.H(Helen). 2010. CrossCheck: An effective tool for detecting plagiarism. *Learned Publishing* 23(1): 9–14.
9. Zhang, Y.H., and X.Y. Jia. 2012. A survey on the use of CrossCheck in detecting plagiarism in scholarly journal articles. *Learned Publishing* 25(4): 292–307.

10. Bowyer, K.W. 2012. Mentoring advice on 'conferences versus journals' for CSE faculty. http://nd.edu/~kwb/Mentoring_Conferences_Journals.pdf (August 2012).
11. Campbell, B. 2012. Conference paper selectivity and impact: A brief overview. Wiley-Blackwell Publishing News (September 2010).
12. Freyne, J., L. Coyle, B. Smyth, and P. Cunningham. 2010. Relative status of journal and conference publications in computer science. *Communications of the ACM* 53(11): 124–132.
13. Menczer, F. 2009. Abolish conference proceedings. *Communications of the ACM* 52(11): 6.
14. Vardi, M. 2009. Conferences vs. journals in computing research. *Communications of the ACM* 52(5): 5.
15. Zhang, L., and W. Glanzel. 2012. Proceeding papers in journals versus the 'regular' journal publications. *Journal of Informetrics* 6: 88–96.
16. Colebatch, C. 2012. Copyright on the conference circuit. http://www.brocku.ca/webfm_send/22504.
17. Elsevier—Artificial Intelligence Journal Editorial Website. Submitting an article to the Artificial Intelligence Journal—Is a conference paper eligible for publication in the AIJ? http://www.ida.liu.se/ext/aijd/authors.html.
18. COPE. 2000. Duplicate publication based on conference proceedings. Case No. 00-20. http://publicationethics.org/case/duplicate-publication-based-conference-proceedings.
19. Meddings, K. 2010. Credit where credit's due: Plagiarism screening in scholarly publishing. *Learned Publishing* 23(1): 5–8.
20. Elsevier. For editors: Questions and answers on policies. http://www.elsevier.com/editors/publishing-ethics/perk/questions-and-answers#policies.
21. Zhang YH(Helen)., HF. Lin, and Q. Ye. 2015. Anti-plagiarism policy of *JZUS-A/B&FITEE*. http://www.zju.edu.cn/jzus/Policy.php.

Part III
What to Do About It

Chapter 7
Promoting Awareness of Publication Ethics

Irene Hames, an experienced journal managing editor, in her book 'Peer Review and Manuscript Management in Scientific Journals', says: 'education can play a large part in decreasing the incidence of misconduct, by making individuals realise that certain behaviour is inappropriate, by creating an awareness of ethical issues, and by introducing the concept of good practice in scientific research and publication from an early age' [1]. As outlined in the following chapter, there are now many examples of policies, rules and guidelines about research and publication ethics; however, policies and rules are not the same as education.

Education, of course, begins in childhood—and so does plagiarism. It is now all too tempting for school children to cut and paste material from the Internet rather than writing 100 % original assignments (although tools such as TurnItIn are designed to detect such plagiarism). Children are taught from their earliest years that both stealing and lying are wrong; schools (and indeed parents) need to help them to understand that plagiarism—taking other people's work and pretending that it is your own—is itself both stealing and lying. And university courses should ideally always include a module on academic, research and publication ethics which makes it absolutely clear that borrowing others' work—whether words, non-textual items or indeed ideas—and presenting it as if it was one's own original work is always plagiarism and will never be permitted. Students should learn that the source of borrowed work must always be acknowledged, with a full citation to the original source so that readers can examine the original in context; and that actual quotations must always be clearly identified typographically, whether by the use of quotation marks (for shorter extracts) or by indented paragraphs (for longer pieces). In addition, they need to understand that copyright material—whether text or non-text—must never be reproduced (other than in very short extracts) without first obtaining the permission of the copyright owner (which may well be the publisher rather than the author).

All human activity may involve misconduct, and thus ethical education has always been an issue of global importance. The particular ethical issue of academic plagiarism is also a global problem. In the more developed countries, the USA established the Office of Scientific Integrity and the Office of Scientific Integrity Review in 1989; in 1999 these two bodies were consolidated into the Office of Research Integrity (https://ori.hhs.gov). The ORI provides, among other policies,

Y.H. Zhang, *Against Plagiarism*, Qualitative and Quantitative Analysis of Scientific and Scholarly Communication,
DOI 10.1007/978-3-319-24160-9_7

one on plagiarism (https://ori.hhs.gov/ori-policy-plagiarism) [2]. The Australian Academic Integrity Standards Project 2010–2012 (http://www.aisp.apfei.edu.au/) [3] aimed 'to develop a shared understanding across the Australian higher education sector of academic integrity standards with the aim of improving the alignment of academic integrity policies and their implementation'. Recently Japan Society for the Promotion of Science Editing Committee (JSPSEC) just issued a book—'For the Sound Development of Science—The attitude of a Conscientious Scientist' [4] in order for scientific researcher and authors to know what is a responsible research activity?

Among less developed countries, China established the Office of Scientific Research Integrity Construction in 2007. However, other countries have yet to follow China's lead [5, 6]. 'China has created the Office of Scientific Research Integrity Construction and begun a comprehensive response to research misconduct, but most low- and middle-income countries have yet to mount a response' [4].

We have investigated in detail the progress of awareness of issues to do with academic and publication ethics in China, to illustrate how the issue can be addressed at a national level.

An Example: China

In Jan 18, 2007, the Ministry of Science and Technology of the People's Republic of China published a key document, 'The construction of Research Integrity' (in Chinese) (http://www.most.gov.cn/kycxjs/). This conveyed to the whole research community the importance of academic research and publication integrity, including the avoidance of plagiarism.

The consequences of this government initiative have been far-reaching. In April–May 2015, we accessed all available information posted online by the Chinese Tier-1 universities and research institutes, as well as the key research management agencies, on their research integrity policies. We also obtained data from CNKI (China National Knowledge Infrastructure) about the number and geographical distribution of Chinese-language journals which use AMLC (Academic Manuscript Literature Checking) system (tool) launched by Tongfang Knowledge Network Technology in 2008, and braced by mass and rich full-texts as the comparing databases which include all contents of China Integrated Knowledge Resources Database about 90 million foreign literature and more than 3.3 billion network resources [7].

Figures 7.1, 7.2, 7.3, 7.4 and 7.5 illustrate the rapid growth in awareness of these issues in China. What this shows is that there has been a rapid change in China, thanks to a top-down approach to the issue of research integrity, cascading from the government itself, through the key research management agencies to individual universities and institutes, and thence to individual journals and journal publishers. Research integrity policies have been created and publicized and research integrity

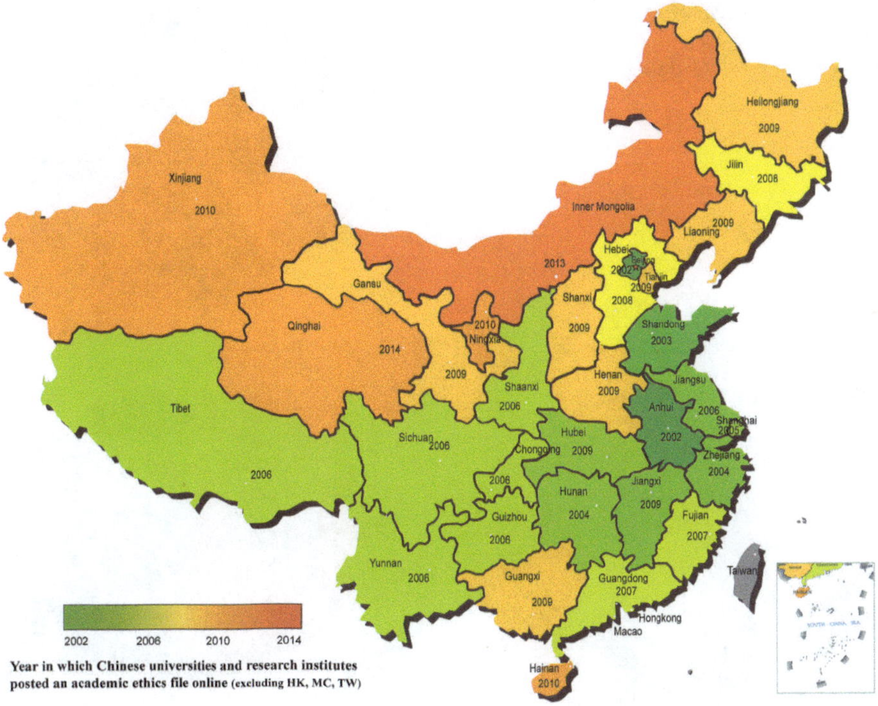

Fig. 7.1 Academic ethics policies and research integrity offices in China

- Chinese Academy of Sciences:
  Moral Construction Committee (Sept. 1996)
- Chinese Academy of Engineering:
  Moral Construction Committee (Aug. 1997)
- National Natural Science Foundation of China:
  Oversight Committee (Nov. 1998)
- Ministry of Education of the People's Republic of China:
  Academic Integrity Committee of Social Science of Council of the Ministry of Education (May 2006)
- Ministry of Science and Technology of the P. R. China:
  Joint Committee on Scientific Integrity (Mar. 2007)

* All of the key agencies are located in Beijing

Fig. 7.2 Establishment of offices of research integrity by key scientific research management agencies in China

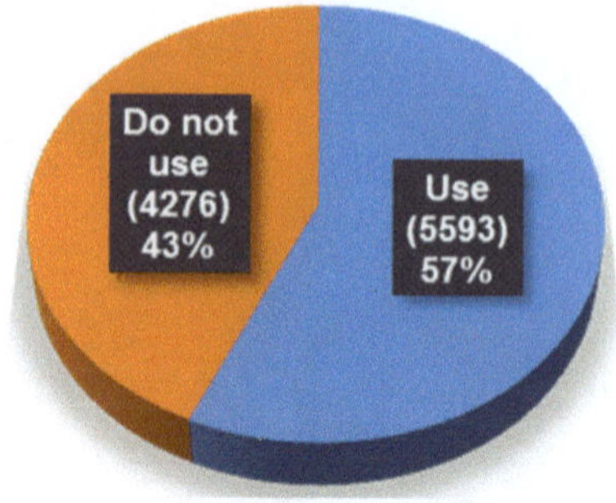

Fig. 7.3 Percentage of Chinese (English-language) journals using CrossCheck

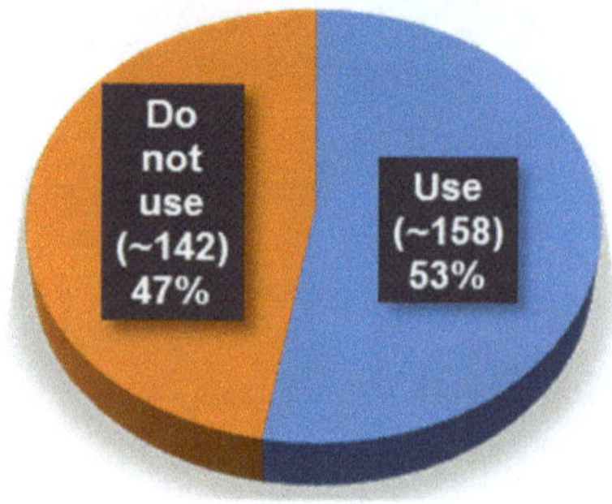

Fig. 7.4 Percentage of Chinese (Chinese-language) journals using AMLC

offices or committees established, and more and more journals have started to use plagiarism detection tools.

From 2002 onwards, almost all of the Tier-1 universities and research institutes in China have now posted an academic ethics policy online. The first universities and institutes to post such policies were in the areas where most of the important research institutes are located, such as Beijing and Shanghai (see Fig. 7.1). Some of them, as well as the key research management agencies, have also established a research integrity office or committee (see Fig. 7.2).

In addition, from 2008 onwards, many Chinese-published journals have taken steps to promote awareness of the importance of research integrity. To date, five journals, *JZUS-A, JZUS-B, Frontiers of Information Technology & Electronic Engineering (FITEE, former JZUS-C), Science Bulletin* and *Cell Research* have posted online their own ethical policies [8–10].

There are about 300 English-language journals published in China; of these, approximately 158 (53 %) are currently using CrossCheck (see Fig. 7.3) to check submitted articles for potential plagiarism. And there are currently 9869 Chinese-language journals published in China, of which 5593 (57 %) are currently using AMLC (Academic Manuscript Literature Checking) from CNKI (China National Knowledge Infrastructure) (see Figs. 7.4 and 7.5).

Fig. 7.5 Percentage (by province) of journals which use AMLC checking in mainland China

We believe that the example of China demonstrates clearly how a national approach to promoting awareness of the importance of ethical issues in research and its publication can play a vital part in raising standards of author and journal behaviour.

References

1. Hames, Irena I. 2007. *Peer review and manuscript management in scientific journals—guidelines for good practice*, 293. Hoboken: Blackwell Publishing.
2. ORI policy on plagiarism https://ori.hhs.gov/ori-policy-plagiarism.
3. The academic integrity standards project: aligning policy and practice in Australian Universities http://www.aisp.apfei.edu.au.
4. JSPSEC. 2015. *For the sound development of science–the attitude of a conscientious scientist*, 157. Tokyo: Maruzen Publishing.
5. Bohannon, J. 2014. Study of massive preprint archive hints at the geography of plagiarism. *Science*. Available from http://news.sciencemag.org/scientific-community/2014/12/study-massive-preprint-archive-hints-geography-plagiarism (accessed December 11, 2014).
6. Ana, J., T. Koehlmoos, R. Smith, and L.L. Yan. 2013. Research misconduct in low- and middle-income countries. *PLoS Med* 10(3): e1001315. doi:10.1371/journal.pmed.1001315.

7. National research integrity seminar held in Beijing (in Chinese) http://www.most.gov.cn/kycxjs/kycxgzdt/200810/t20081029_64662.htm.
8. Anti-Plagiarism policy of *JZUS-A/B&FITEE* http://www.zju.edu.cn/jzus/Policy.php.
9. Science China life sciences: Instructions for authors (website edition)-ethical responsibilities of authors http://life.scichina.com:8082/sciCe/EN/column/column13292.shtml.
10. Cell research guidelines: For author: Ethical considerations http://www.cell-research.com/guidelines.asp.

Chapter 8
Avoiding Plagiarism as an Author

Principles and Responsibilities

As an author in order to avoid plagiarism, the best way is learning and compliance with the basic principles of good academic from your academic career's starting. And as signaled by the University of Oxford's website 'avoiding plagiarism is not simply a matter of making sure your references are all correct, or changing enough words so the examiner will not notice your paraphrase; it is about deploying your academic skills to make your work as good as it can be' [1]. Further from the perspective of the global scientific integrity, as points by a new book edited by Japan Society for Promotion of Science Editing Committee, 'Scientists should themselves take positive steps to learn anew the nature of principled research and to build upon that concept in fostering the next generation of scientists so as to soundly advance science and establish public trust in it [2]'. And in Singapore Statement on Research Integrity, [3] there are total 14 principles and professional responsibilities among which Article 7 requires 'Researchers should acknowledge in publications the names and roles of those who made significant contributions to the research, including writers, funders, sponsors, and others, but do not meet authorship criteria'. We can know these rules are fundamental to the integrity of research wherever it is undertaken. In other words, If you are a honest scientific author, you have to fulfill the responsibilities.

University Examples for Author Conduct Rules

In recent years, a growing number of leading universities have provided their own rules of conduct for scholarly authors, whether undergraduates, postgraduates or faculty. Some outstanding examples are listed below.

Y.H. Zhang, *Against Plagiarism*, Qualitative and Quantitative Analysis
of Scientific and Scholarly Communication,
DOI 10.1007/978-3-319-24160-9_8

USA

- **Cornell University**

Code of Academic Integrity (http://cuinfo.cornell.edu/aic.cfm)

- **Harvard University**

Harvard Guide to Using Sources (http://usingsources.fas.harvard.edu/icb/icb.do)

- **Harvard Medical School**

White Paper: Plagiarism and Research Misconduct (2010) (http://hms.harvard.edu/sites/default/files/assets/About_Us/COI/files/plagiarism_statement_121510.pdf)

- **Massachusetts Institute of Technology (MIT)**

Academic Integrity at the Massachusetts Institute of Technology: a handbook for students (2012) (http://web.mit.edu/academicintegrity/handbook/handbook.pdf)

- **Princeton University**

Academic Integrity (2011) (http://www.princeton.edu/pr/pub/integrity/pages/intro/)

- **Yale University**

What is Plagiarism? (http://writing.yalecollege.yale.edu/advice-students/using-sources/understanding-and-avoiding-plagiarism/what-plagiarism)

UK

- **University of Cambridge**

University-wide statement on plagiarism (http://www.admin.cam.ac.uk/univ/plagiarism/students/statement.html)

- **University of Oxford**

Plagiarism (http://www.ox.ac.uk/students/academic/guidance/skills/plagiarism)

Japan

- **University of Tokyo**

Code of Research Integrity (2006, in Japanese) (http://www.u-tokyo.ac.jp/ja/administration/codeofconduct/index.html)

China

- **Zhejiang University**

Academic ethical code and management policy (2009-No. 15—Document, in Chinese) (http://xfjs.zju.edu.cn/redir.php?catalog_id=5&object_id=346)

Australia

- **Australian National University**

Academic Misconduct Rules 2014 (http://www.comlaw.gov.au/Details/F2014L01785)

Journal Publisher's Guideline Example

Some publishers also provide guidance on their websites for authors; for example, Elsevier offers educational materials for authors on 'Ethics in Research and Publication' (http://www.elsevier.com/ethics). And Wiley-Blackwell has its 'Best Practice Guidelines on Publication Ethics—a Publisher's Perspective'(https://authorservices.wiley.com/bauthor/publicationethics.asp) [4] in English, Chinese and Japanese, respectively. Particularly based on its "Author Services", authors will learn not only what are both ethical broad ethical issues and practical points, but also be aware of principle of transparency that requires authors must let the reader know who did the work and has the work been published before in order to avoid plagiarism and promote research integrity.

In 2015, *JZUS* became the first Chinese journal to post an Anti-Plagiarism Policy (http://www.zju.edu.cn/jzus/Policy.php) that will be our journal guidelines for our authors, which clear point.

The following are acceptable, provided always that (a) the quotation (if any) is typographically identified (by quotation marks or, for longer extracts, indentation), (b) the source is acknowledged in the text, and (c) a full citation to the original is given:

(1) Quotation of a modest amount (under 100 words) [5] of the author's own or others' text;
(2) Paraphrase of previously published text in the author's own words;
(3) Repetition of someone else's ideas;
(4) Reproduction of a chart, image, table or key equation from your own or someone else's work (provided copyright permission has been obtained from the original copyright owner, and acknowledgement is included in whatever form they request);

(5) In Biosciences papers it is acceptable to reproduce the description of a standard/homemade method from a previously published source, provided the source is properly acknowledged;
(6) Republication of a previously published conference paper is acceptable, if 60 % or more of the content is new and substantive (provided copyright permission has been obtained from the original copyright owner, and acknowledgement is included in whatever form they request).
(7) Republication of a paper in translation is acceptable only if, in the view of the peer reviewers, it is necessary in order to reach part of the intended readership, and then only with copyright permission from the original publisher.

The following are unacceptable in any circumstances:

(8) Duplicate publication of an entire article;
(9) Major plagiarism of the work of others (SMSI > 10 % or OSI > 35 %);[1]
(10) Serious self (or team)-plagiarism (SMSI > 10 % or OSI > 35 %);
(11) Review papers which reproduce substantial amounts of the texts discussed (OSI > 35 %) (See footnote 1).

The key principle which authors should follow is that of integrity: Ralph Waldo Emerson wrote in the 19th century, 'Nothing is at last sacred but the integrity of your own mind' [6].

Basic Principles to Ensure that Authors Avoid Plagiarism

Other authors' work should always be treated with the same respect that you would wish for your own work. That means:

- Not relying excessively on quoted (or your own previously published) material; if you have nothing new to add, don't write it!
- Always identifying the source of ideas, words, data, figures etc., with a full citation to the originally published source, whether the work is your own or someone else's. Reference styles differ from journal to journal, but are generally variants of the 'Vancouver' style (common in science, technology and medical journals) or the 'Harvard' style (more usually found in social sciences and humanities journals) [7, 8].

[1]OSI: the **Overall Similarity Index** represents the 'percentage of similarity between a submission and information existing in the Crosscheck/iThenticate databases selected as search targets' [minor: <25 %; middle: 25–35 %; major: >35 %]; SMSI: the **Single Match Similarity Index** represents the percentage of similarity from a single source [minor: <6 %; middle: 6–10 %; major: >10 %]

- Always identifying actual quotations from someone else's work with quotation marks (if the extract is short) or indentation (if it is long). Bear in mind that copyright issues are likely to arise if you are quoting substantial passages (or non-textual material, such as figures, graphs or diagrams) from anyone else's work; you will need to seek the original publisher's permission (the author's permission alone is not enough). Even if you reword, rephrase, summarize or translate someone else's writing (or your own previously published) material, you should still credit the original source.

References

1. Oxford Student: Plagiarism. http://www.ox.ac.uk/students/academic/guidance/skills/plagiarism.
2. For the Sound Development of Science. 2015. Ed. Japan Society for Promotion of Science Editing Committee. Maruzen Publishing.
3. Singapore Statement on Research Integrity. http://www.singaporestatement.org.
4. Graf, C., E. Wager, A. Bowman et al. 2007. Best practice guidelines on publication ethics: A publisher's perspective. *International Journal Clinical Practice* 61(Suppl. 152): 1–26.
5. Wager, L. 2011. How should editors respond to plagiarism? COPE discussion paper. Available from http://publicationethics.org/files/COPE_plagiarism_disc.doc_26Apr11.pdf.
6. Penders, Bart. 2012. Commence to research ethics: Zero tolerance. http://www.nature.com/news/research-ethics-zero-tolerance-1.9756?WT.ec_id=NEWS-20120117 (This link doesn't work).
7. The Chicago Manual of Style. The essential guide for writers, editors and publishers (14th Edn.). The University of Chicago Press. www.press.chicago.edu.
8. Referencing Style Guides. https://www.library.uq.edu.au/help/referencing-style-guides.

Chapter 9
Detecting Potential Plagiarism

There are now a variety of online tools available which can compare a submitted text with texts already in its database (or freely available online). The most widely used are CrossCheck, which is mainly used for texts in English, and Academic Manuscript Literature Checking (AMLC) from China National Knowledge Infrastructure (CNKI) that almost covers all contents of Chinese Integrated Knowledge Resources Database and is mainly used for texts in Chinese. These tools enable journal editors to check submitted material against large databases of already published content, and to produce succinct statistical reports of the amount of similar or identical wording found. The author's editorial team relies on CrossCheck, which is described in detail in this chapter, to carry out these checks.

While a high score from one of these tools does not necessarily prove plagiarism (there are special cases, discussed below, in which verbatim repetition may be perfectly acceptable), it should alert the journal editor to the possibility, and encourage him or her to investigate more closely.

At the same time, the absence of a suspiciously high score does not absolutely prove the absence of plagiarism—ideas, as well as words, are subject to plagiarism; in addition, none of these tools can—as yet—search for duplication in non-textual content, such as tables and figures, or in different languages. So the editor and his or her reviewers still have the responsibility of checking, in the light of their knowledge of the field, for any content which seems familiar.

CrossCheck

CrossCheck powered by iThenticate is an initiative started by CrossRef [1], or PILA-the Publisher International Linking Association that is an association of scholarly publishers that develops shared infrastructure to support more effective scholarly communications, because in 2006 the Board of the CrossRef project raised plagiarism as an area of particular concern. As a result, the CrossCheck plagiarism detection project (http://www.crossref.org/crosscheck/index.html) was piloted with seven international publishers and a technology partner (iParadigms)

Y.H. Zhang, *Against Plagiarism*, Qualitative and Quantitative Analysis of Scientific and Scholarly Communication,
DOI 10.1007/978-3-319-24160-9_9

during late 2007 and early 2008. CrossCheck was commercially launched in June 2008, and won the ALPSP Award for Innovation in the same year [2].

CrossCheck is an international project intended to help cope with the high incidence of plagiarism in recent years, by reliably detecting the extent of duplication between texts. It is led by its parent organization, CrossRef, and many global publishing groups are members.

As CrossCheck's workflow shows (Fig. 9.1), its checking process relies mainly on two components. The first is the content database—a huge and constantly growing repository of content deposited by CrossRef's more than 600 publisher members and by scholarly databases such as PubMed, arXiv, EBSCO, Gale etc. As mentioned in Chap. 1, CrossCheck's underlying database now includes more than 126,260 journal titles and over 42 million individual content items from over 642 publishers, and more than 190,000 additional manuscripts are uploaded every month [3].

The second key component is the iThenticate software; this compares submitted manuscripts with the content database, and produces reports that can help editors to verify originality [4–7]. The tool scans and detects duplicated text, thus helping to identify possible plagiarism. Although CrossCheck itself cannot actually identify plagiarism, it can direct journal editors' attention to the possibility of plagiarism. It does this through its similarity reports which provide two figures: the Overall Similarity Index (OSI) and the Single Match Similarity Index (SMSI) [6, 8]. The OSI represents the 'percentage of similarity between a submission and information existing in the iThenticate databases selected as search targets', while the SMSI represents the percentage of similarity from a single source. A high score on either or both indices may suggest duplicate publication, self-plagiarism or

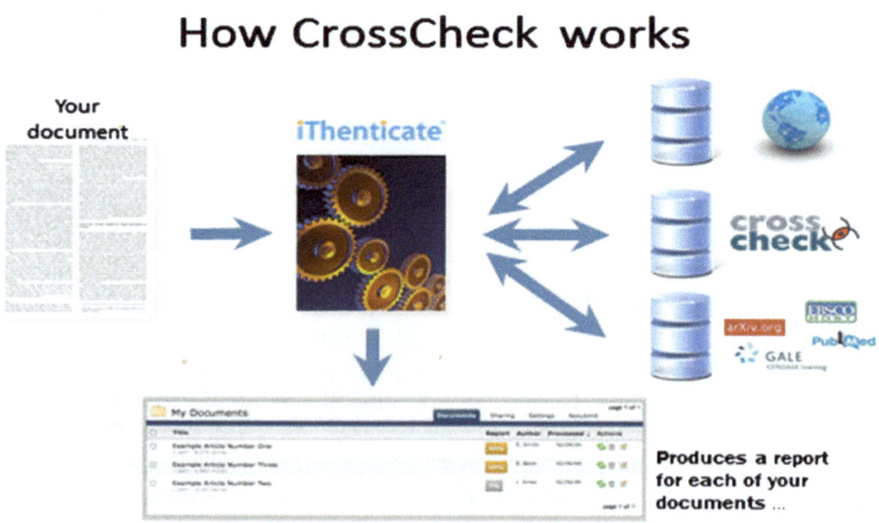

Fig. 9.1 How CrossCheck works [3] (reproduced with permission of CrossRef)

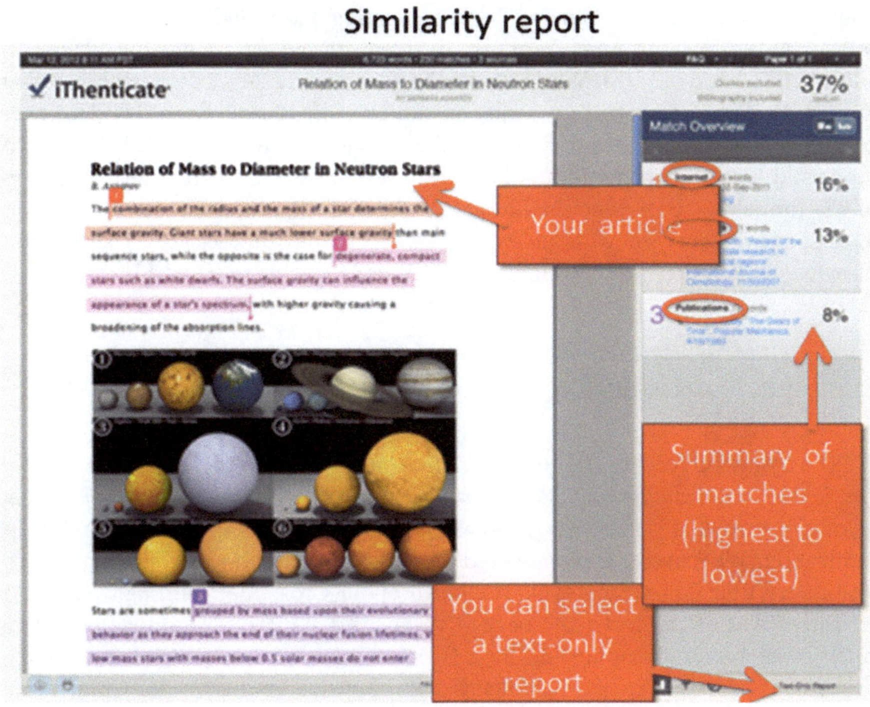

Fig. 9.2 A CrossCheck similarity report [3] (reproduced with permission of CrossRef)

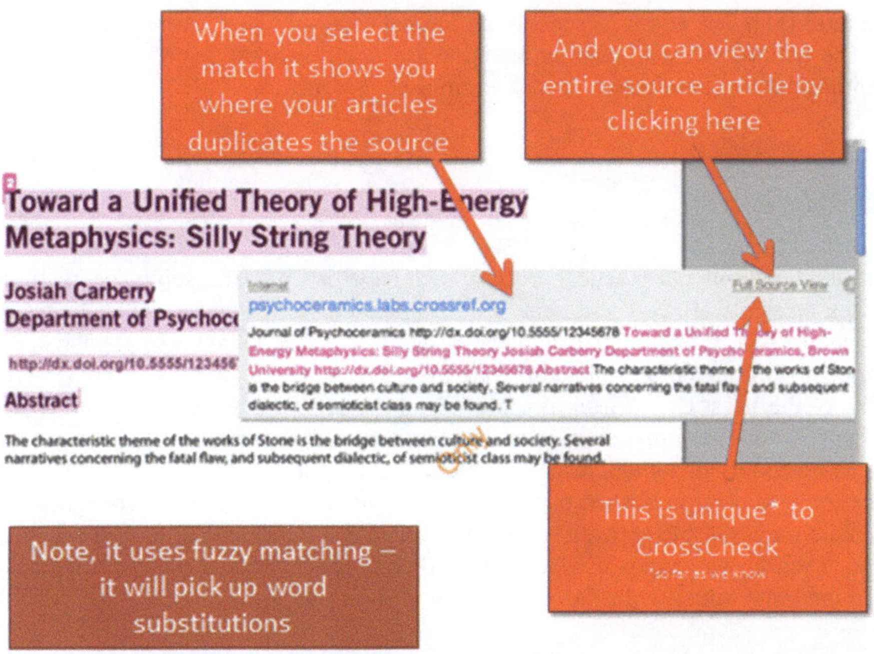

Fig. 9.3 How to deal with a similarity report [3] (reproduced with permission of CrossRef)

plagiarism of the text of others. By investigating these and exercising editorial judgement, editors can much more easily identify and deal with true plagiarism (Figs. 9.1, 9.2 and 9.3) [3]. While at first the similarity reports can appear daunting, 'people do start to get a feel for what constitutes a significant match' [5]. However, CrossCheck and other 'Plagiarism-checking' systems can only identify similar or duplicated text; they cannot identify ideas, translated text, figures, image, graphs, tables, formulae, all of which might also consist of, or contain, content which has been plagiarized/copied without attribution. To identify these, the expert subject knowledge of the peer reviewers and the editor is always required [3].

Experience of Using CrossCheck at *JZUS*

As Managing Editor [9, 10], of *Zhejiang University-Science A/B/C (A: Applied Physics, B: Biomedicine and Biotechnology, C: Computers and Electronics)* I had become aware that some peer reviewers were commenting on the problem of duplication. Whenever we encountered this problem, verifying the suspect content was a very time-consuming task. We heard about CrossCheck through our membership of ALPSP, when CrossCheck won the 2008 ALPSP Award for Innovation; in the same year, our application for a Special Grant for a Key Academic Journal project was approved by the National Natural Science Foundation of China. Soon afterwards, the *Journal of Zhejiang University Science (JZUS)* became the first member of CrossCheck in China. Since then, we have used CrossCheck as part of our standard review process for submitted papers—we call it the 'third eye of editors' [11]. The author described the journal's use of CrossCheck in 2010 in an article entitled 'CrossCheck: an effective tool for detecting plagiarism', which is reproduced verbatim below (with the publisher's permission) [7].

Introduction

Generally, during our practical process, each paper is CrossChecked twice: the first check takes place before it is sent to international reviewers; a second check takes place just before 'online-first' publication, to ensure that no potential plagiarism is missed owing to the inevitable time-lag in updating the CrossCheck database. The date of the latest CrossCheck is included on the first page for each journal paper (Fig. 9.4) for the information of readers, authors, and databases.

The majority of authors behave correctly, submitting papers that bear little or no similarity to other published papers. However, around 22.8 % of papers appear to contain unreasonable copying or self-plagiarism, and about a quarter of these give rise to serious suspicions of plagiarism and copyright infringement; in some cases, the similarity with the plagiarized original was as high as 83 %.

Shen et al. / J Zhejiang Univ Sci B 2009 10(1):1-6 1

Journal of Zhejiang University SCIENCE B
ISSN 1673-1581 (Print); ISSN 1862-1783 (Online)
www.zju.edu.cn/jzus; www.springerlink.com
E-mail: jzus@zju.edu.cn

Estrogen receptor expression in adrenocortical carcinoma

Xiao-cao SHEN[1], Cai-xiao GU[1], Yi-qing QIU[†‡1], Chuan-jun DU[1], Yan-biao FU[2], Jian-jun WU[3]
([1]Urology Department; [2]Pathology Department; [3]Radiology Department, the Second Affiliated Hospital, School of Medicine, Zhejiang University, Hangzhou 310009, China)
[†]E-mail: q_yiqing@yahoo.com
Received Mar. 6, 2008; Revision accepted July 25, 2008; Crosschecked Dec. 9, 2008

Abstract: Objective: Adrenocortical carcinoma (ACC) is a rare but highly malignant tumor, and its diagnosis is mostly delayed

Fig. 9.4 The date of the most recent CrossCheck is indicated on the first page of each journal paper

Four Distinct Types of Plagiarism

Four distinct types of plagiarism were identified, which we consider sufficiently serious to be considered as a form of academic misconduct:
Duplicate publication
Self- (or team) plagiarism
Direct copying of Methods section, with new data inserted
Uncited or excessive extracts.
A fuller report of the findings, in Chinese, was published in *ScienceTimes* [11].

Duplicate Publication

CrossChecking identified the fact that a few authors had contributed almost identical papers to several journals, or had submitted—completely unchanged—papers previously published in conference proceedings or electronic journals. If the similarity is more than 40–50 %, we automatically reject the article on the basis of duplication. For examples, in May, 2009 the last CrossCheck identified one article which duplicated about 78 % of the content from a paper by the same author published in an IEEE journal in early 2009 (Fig. 9.5).

Identification of duplicated text is not difficult using CrossCheck. However, currently CrossCheck is unable to check duplication in figures and tables, so we have recourse to other sources (Google, PubMed Central etc.) for further analysis of articles highlighted by CrossCheck. Comparison of a paper from France and one

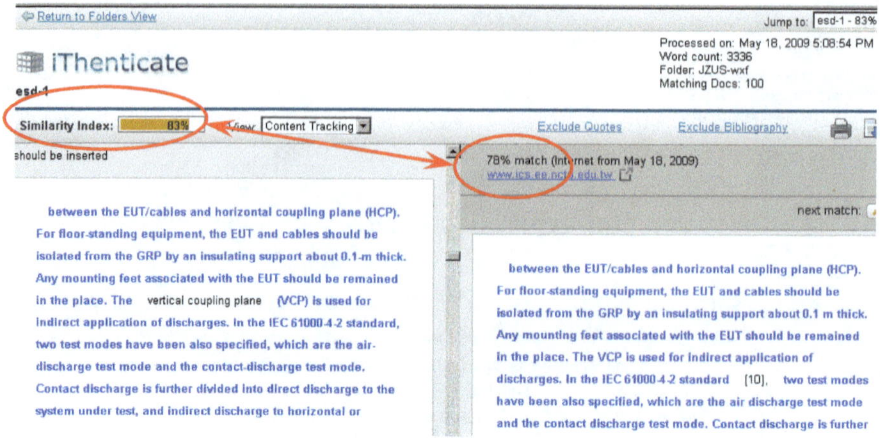

Fig. 9.5 Duplicate publication

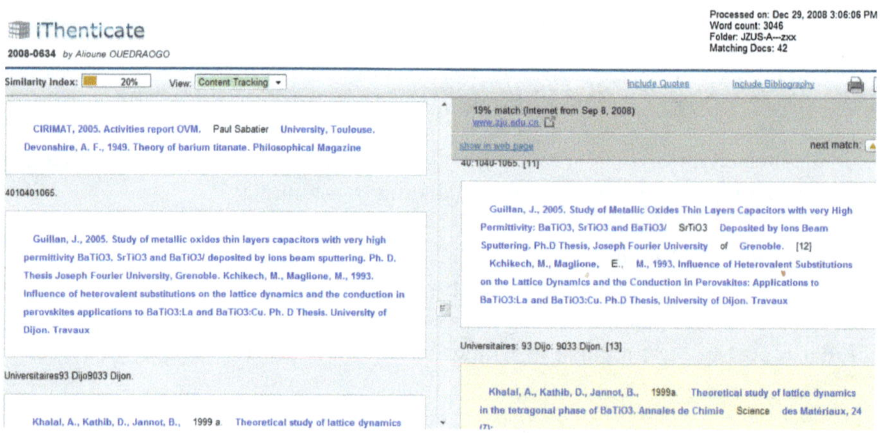

Fig. 9.6 Identical references in two apparently different papers

from Burkina Faso showed that, while only 18 % of the main text was duplicated, the references were identical (Fig. 9.6). When we referred to PubMed Central's full-text database, we discovered that the figures and tables were completely duplicated from the earlier publication, so that the actual duplication was nearer 80 %. This shows that our editors cannot rely on CrossCheck alone, but also have to make additional efforts to detect duplication at the time of submission and of publication.

In another typical example, a paper was found to have the same abstract as another previously published paper. Further investigation revealed that the author's PhD thesis had already been published online through a university press, and that the author had also already published, before PhD graduation, two papers containing the core content of his thesis. However, he argued that although 39 % of the paper replicated previous publications, the rest of the material was previously unpublished content from his PhD thesis, and thus that the whole paper should still be published. Since the whole thesis was already available online, and its core content had been published 5 years previously, we felt that it was unacceptable to republish it unless it contained significant new information, and the article was therefore rejected.

In our view, duplicate publication injures the interests of many journals, wastes publication resources and should be condemned both by academia and the publishing industry. As Errami and Garner [12] state, 'the repeated publication of the same results by those who conducted the research is ethically questionable. It not only artificially inflates an author's publication record but places an undue burden on journal editors and reviewers, and is expressly forbidden by most journal copyright rules'. Further work is needed to define relevant criteria.

Self- (or Team) Plagiarism

Another familiar phenomenon is self-plagiarism (or plagiarism of the publications of other team members). This can frequently be found in papers of authors from the same research program (Fig. 9.7). Some authors, or even program leaders, believe that this is justified by different focuses in the same research project, even when the equipment and methods adopted are the same; thus they do not feel it is unreasonable to duplicate parts of the introduction, methods and discussion sections.

However, in our view, once a paper is published the authors should not recycle any of its content in new papers. Self-plagiarism wastes not only the publication resources of journals but also the time of readers. Instead, authors should simply cite previous studies, giving no more than an overview in their current paper. It is

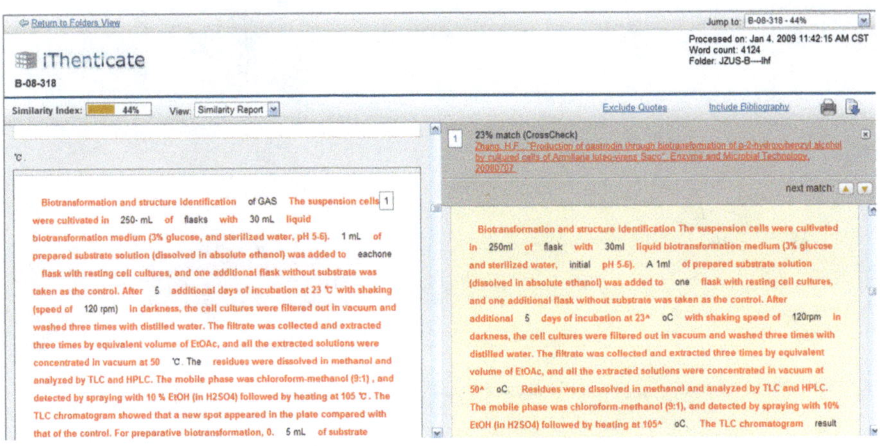

Fig. 9.7 Plagiarism of the work of members of the same research team

preferable to put combine the content of several papers together to form a one high quality paper, rather than repeating some of the contents to form different papers.

Publishers also object to this practice. As Arnout Jacobs, vice-president of the science and technology department, Chinese section, at Elsevier, says: ‘it is a nuisance for journal editors when researchers publish a series of highly similar papers. Often, these papers could easily be rewritten as one single excellent paper. This happens less often in the USA or Europe, where editors or funding agencies check earlier publications routinely as a reference and authors would be judged negatively for publishing multiple papers with the same topics and replicated contents’ [13].

Direct Copying of Methods Section, with New Data Inserted

This is a particularly common phenomenon in biomedical papers, where all or part of the Methods section may be copied verbatim, only changing some of the experimental conditions and data (see Fig. 9.8). Some authors feel that it is a acceptable to copy all or part of the Methods section from a previously published article, simply inserting their own data.

However, we observe that this type of direct copying is hardly ever found in leading journals such as *Science* and *Nature*. In principle we believe that, although much research refers to or repeats others’ successful methods in testing new materials and discussing new results, the authors should use their own language to describe and summarize their methods and ideas.

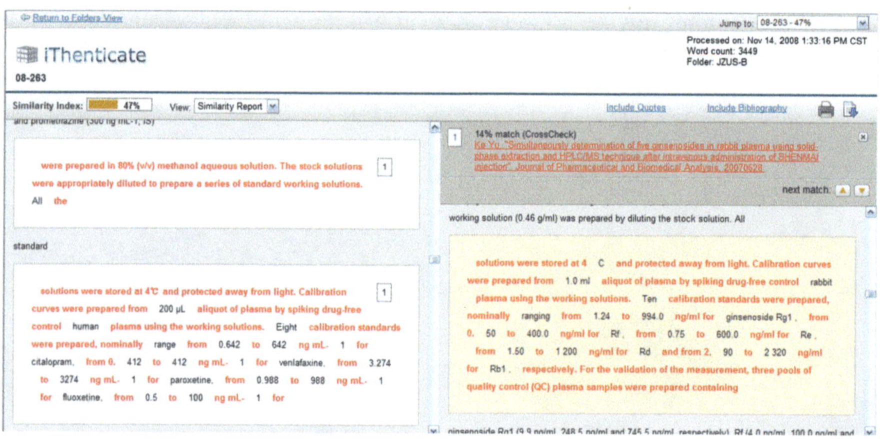

Fig. 9.8 Direct copying of methods section with new data inserted

Uncited or Excessive Extracts

We found that some authors incorporated extracts from other papers without providing citation details (see Fig. 9.9). In one instance, when we raised the matter with the author, he argued that, since his own view was identical to that of the other author, it was acceptable to use the same words without citation. However, such conduct misleads readers into believing that they are reading the author's own words and, quite apart from its academic impropriety, this is an infringement of copyright.

Sometimes, too, authors believe that, with a full citation, it is reasonable to copy whole paragraphs from other papers; this is not the case, and the 'fair dealing' rules always apply.

The phenomenon of 'copy and paste' is also all too common, particularly in papers from non English-speaking authors. In a few extreme cases, we found that many sentences and whole paragraphs were identical to those in published papers, and scarcely any of the words were the authors' own (see Fig. 9.10).

The Council of Science Editors gives clear definitions of piracy and plagiarism [14].

Piracy is defined as the appropriation of ideas, data, or methods from others without adequate permission or acknowledgment. Again, deceit plays a central role in this form of misconduct. The intent of the perpetrator is the untruthful portrayal of the ideas or methods as his or her own.

Plagiarism is a form of piracy that involves the use of text or other items (figures, images, tables) without permission or acknowledgment of the source of these materials. Plagiarism generally involves the use of materials from others, but can apply to researchers' duplication of their own previously published reports without acknowledgment (this is sometimes called self-plagiarism or duplicate publication).

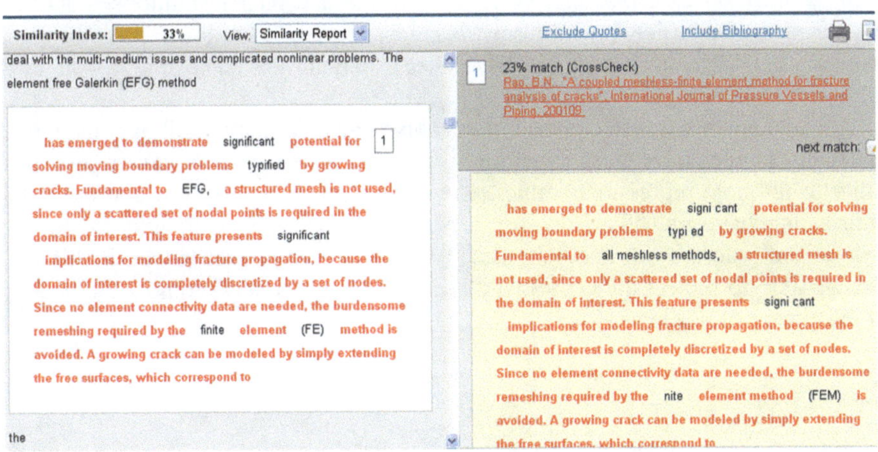

Fig. 9.9 Uncited extracts from other papers

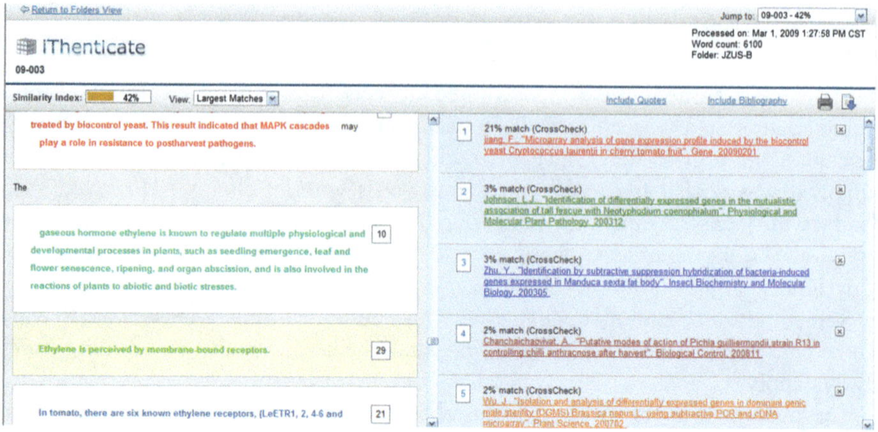

Fig. 9.10 Copy and paste, with almost no original text

Authors should ensure that any article they submit for publication is original and does not contain plagiarized content from either their own or others' work. If an author's text follows the source so closely that the result is more of the quotation than a paraphrase, it constitutes plagiarism; the author must either completely recast the summary in his or her own words (changing a few words is not sufficient), or quote explicitly.

Conclusions

The importance of science should be measured by the quality of papers rather than their quantity. In China, as elsewhere, researchers and their institutions should be evaluated on the basis of real original research results, rather than on the basis of paper output. An emphasis on quantity rather than quality is liable to lead to authors taking short cuts such as plagiarism.

Academia is not a perfect world; inevitably academic journals all over the world are likely to encounter these or similar problems. As editors, we have a responsibility to promote professional ethics. CrossCheck enables us to see that most scientists do behave ethically. However, it is up to the editorial community to propose criteria and processes for handling these types of academic misconduct. In this way we can help to protect the copyrights of original authors, and promote the healthy development of academic journals.

Practical Lessons for Journal Editors

We believe that journal editors have a key role in maintaining and indeed raising standards of academic research and publication ethics, by identifying and correcting plagiarism whenever they encounter it. This task is made immeasurably easier by the availability of computerized tools such as CrossCheck and AMLC. We would strongly recommend that every journal publisher makes use of the tools that are available, while recognizing that expert judgement, on the part of both peer reviewers and editors themselves, will still be essential both to identify forms of plagiarism which cannot be identified by text-matching alone, and to determine the nature and severity of instances of plagiarism when they are encountered.

References

1. CrossCheck in the news. http://www.crossref.org/crosscheck/index.html.
2. ALPSP announces winners of 2008 awards. http://www.crossref.org/01company/pr/ALPSP_Awards_2012_Winners.pdf.
3. Pentz, E. 2015. CrossCheck: 2015 CrossRef workshop in Shanghai, 29 April 2015. http://www.crossref.org/crosscheck/index.html.
4. American Chemical Society (ACS). 2013. American Chemical Society Implements CrossCheck Plagiarism Screening. http://www.acs.org/content/acs/en/pressroom/newsreleases/2013/october/acs-implements-crosscheck-plagiarism-screening.html
5. COPE: Plagiarism screening: An update on CrossCheck. *Ethical* p. 5. (http://publicationethics.org/files/u661/EthicalEditing_Summer2010_final.pdf)
6. iThenticate Admin Webinar. 2015. 19 February 2015 (http://www.slideshare.net/CrossRef/crosscheck-ithenticate-admin-webinar).
7. Zhang, Y.H. 2010. CrossCheck: An effective tool for detecting plagiarism. *Learned Publishing* 23: 9–14.
8. Zhang, Y.H., and X.Y. Jia. 2012. A survey on the use of CrossCheck for detecting plagiarism in journal articles. *Learned Publishing* 25(4): 292–307. doi:10.1087/20120408.
9. ALPSP Alert. 2014. *Missed the ALPSP conference—read the blog (Issue no. 209, October 2014).* UK: Association of Learned, Professional and Society Publishers. http://Alert.alpsp.org.
10. Kavanagh, S. 2014. Innovation versus 'very good' journals. http://blog.alpsp.org/2014_08_01_archive.html.
11. Zhang,Y.H. 2009. Academic misconduct as perceived from recent examples (7 May). *ScienceTimes* (in Chinese). http://www.sciencenet.cn/dz/dznews_photo.aspx?id=6007.
12. Errami, M., and H. Garner. 2008. Commentary—a tale of two citations. *Nature* 451: 397–399. http://dx.doi.org/.
13. Jacobs, A. 2008. How to counter academic dishonesty in STM journals (5 April 2008). *Sciencenet*. www.sciencenet.cn/htmlnews/200842818245120488l.html (in Chinese).
14. http://www.councilscienceeditors.org/editorial_policies/whitepaper/3-1_misconduct.cfm#3.1.3.

Chapter 10
Dealing with Plagiarism as an Editor

As comments by *Nature* in 2010, 'Major science publishers are gearing up to fight plagiarism… including Elsevier and Springer, are set to roll out software across their journals that will scan submitted papers for identical or paraphrased chunks of text that appear in previously published articles [1]'. And since 2011, COPE has been encouraging publishers and journal editors to issue regulations/codes for their own work and reference them or incorporate them into their editorial policies and instructions for authors [2]. It is true that journals should have (and publicise) their own policy for dealing with plagiarism, as well as other forms of ethical misconduct: 'It is a very good idea to have a policy in place for dealing with these issues, so that the journal is not seen to be approaching them in a haphazard or ad hoc way' [3].

Important International Advice Sources

It is not necessary to construct a policy from scratch; there are many sources of expert international advice. Some of the most important are listed below.

The Committee on Publication Ethics (COPE) (http://publicationethics.org)

COPE was established in 1997. Its website provides the definitive source of information and guidelines on international scientific ethical standards and procedures for editors, authors and researchers, covering the wide gamut of types of ethical problems and misconduct. Key documents are the '**Code of Conduct and Best Practice for Journal Editors**' (http://publicationethics.org/files/Code_of_conduct_for_journal_editors.pdf) and the 'Code of Conduct for Journal Publishers' (http://publicationethics.org/files/Code%20of%20conduct%20for%20publishers%20FINAL_1_0.pdf). A range of other guidelines is also available on the COPE website (http://publicationethics.org/resources/guidelines).

Y.H. Zhang, *Against Plagiarism*, Qualitative and Quantitative Analysis of Scientific and Scholarly Communication,
DOI 10.1007/978-3-319-24160-9_10

COPE also organizes regular conferences on research integrity; during the 2nd World Conference on Research Integrity in Singapore in 2010, COPE issued an invaluable series of international standards for responsible research publication for editors and authors (http://publicationethics.org/international-standards-editors-and-authors) [4]. In addition, COPE provides authoritative guidance, including recommended policies and workflows, for publishers and journal editors on aspects of suspected or actual plagiarism, including self-plagiarism (http://publicationethics.org/resources/flowcharts).

COPE also provides a forum for its 9000-plus members (from all academic fields and all parts of the world) to discuss various cases encountered in the field of academic publishing [5]. To date, the COPE website describes no fewer than 57 cases mainly focused on plagiarism—these provide useful examples for researchers, students, authors and editors.

Furthermore, in recent years, it is because of COPE's guidance and encouragement, many academic organizations, publishers, journals and editors have posted or are creating online their own ethical policies, codes of practice for their authors [6]. These initiatives to curb misconduct (or plagiarism) play an effective role.

The European Association of Science Editors (EASE) (http://www.ease.org.uk)

EASE is an association intended for all those 'interested in science communication and editing'. It has some 550 members in many parts of the world. EASE produces a number of key publications and guidelines. The 'Science Editors' Handbook' (http://www.ease.org.uk/publications/science-editors-handbook—not freely available to non-members) contains a substantial section on Ethics. There is also an invaluable 'Toolkit for Journal Editors', including a Publication Ethics Checklist (http://www.ease.org.uk/sites/default/files/ease_guidelines-june2014-ethics.pdf).

World Association of Medical Editors (WAME) (http://www.wame.org)

WAME is 'a global association of editors of peer-reviewed medical journals who seek to foster cooperation and communication among editors, improve editorial standards, promote professionalism in medical editing through education, self-criticism, and self-regulation, and encourage research on the principles and practice of medical editing.'

Its website includes a range of policies and recommendations, as well as ethics resources (http://www.wame.org/policies-and-resources).

Council of Science Editors (CSE) (http://www.councilscienceeditors.org)

'The Council of Science Editors (CSE) is an international membership organization for editorial professionals publishing in the sciences. [Its] purpose is to serve over 800 members in the scientific, scientific publishing, ad information science communities by fostering networking, education, discussion, and exchange. [Its] aim is to be an authoritative resource on current and emerging issues in the communication of scientific information.'

CSE's Publications include a White Paper on Publication Ethics (2012) (http://www.councilscienceeditors.org/resource-library/editorial-policies/white-paper-on-publication-ethics/).

Many individual publishers and journals have also posted online their own policies in order to guide their editors. Some examples are:

Springer 'Publishing Ethics for Journals—A Guide for Editors-in-Chief, Associate Editors, and Managing Editors' (https://www.springer.com/gp/partners/society-zone-issues/springer-s-guide-publishing-ethics-for-journals/15064) in which Springer not only provides practical guidance to Journal Editors, but also strongly recommends journals to develop a publishing ethics policy for their own journal, join COPE and thereby adhere to the principles of COPE, committing to investigate allegations of misconduct and to ensure the integrity of research.

Elsevier Its 'Publishing Ethics Resource Kit (PERK) (http://www.elsevier.com/editors/perk) provides flowcharts to guide editors through processes required to deal with different forms of publishing ethics abuse. Especially by its General Guidelines (all decision trees) and Form Letters, we can learn Relevant Elsevier Policies to deal with Plagiarism complaints in detail (http://www.elsevier.com/editors/publishing-ethics/perk/plagiarism-complaints)

Association of Computing Machinery (ACM) ACM Policy and Procedures on Plagiarism (https://www.acm.org/publications/policies/plagiarism_policy) that includes total 7 principles or steps 1. Definition and Context; 2. Notifying ACM of Alleged Plagiarism; 3. Investigation; 4. Results of an Investigation; 5. Confidentiality; 6. Penalties for Plagiarism; 7. Third-Party Claims against ACM.

Highly detailed practical advice on dealing with plagiarism, and other problems to do with publishing ethics, is also given in some excellent reference works, which we strongly recommend, including the following:

(1) Sally Morris et al. (2013): The Handbook of Journal Publishing, 467 pp. Cambridge University Press.
(2) Irene Hames (2007): Peer Review and Manuscript Management in Scientific Journals—Guidelines for good practice, 293 pp. ALPSP and Blackwell Publishing.

(3) The EASE Science Editors' Handbook, 2nd edition, edited by Pippa Smart, Hervé Maisonneuve and Arjan Polderman. European Association of Science Editors, 2013.

What to Do if You Suspect Plagiarism

Two of the COPE flowcharts mentioned above are focused on what to do when the journal editor encounters the following situations: (a) suspected plagiarism in a submitted manuscript and (b) suspected plagiarism in a published manuscript.

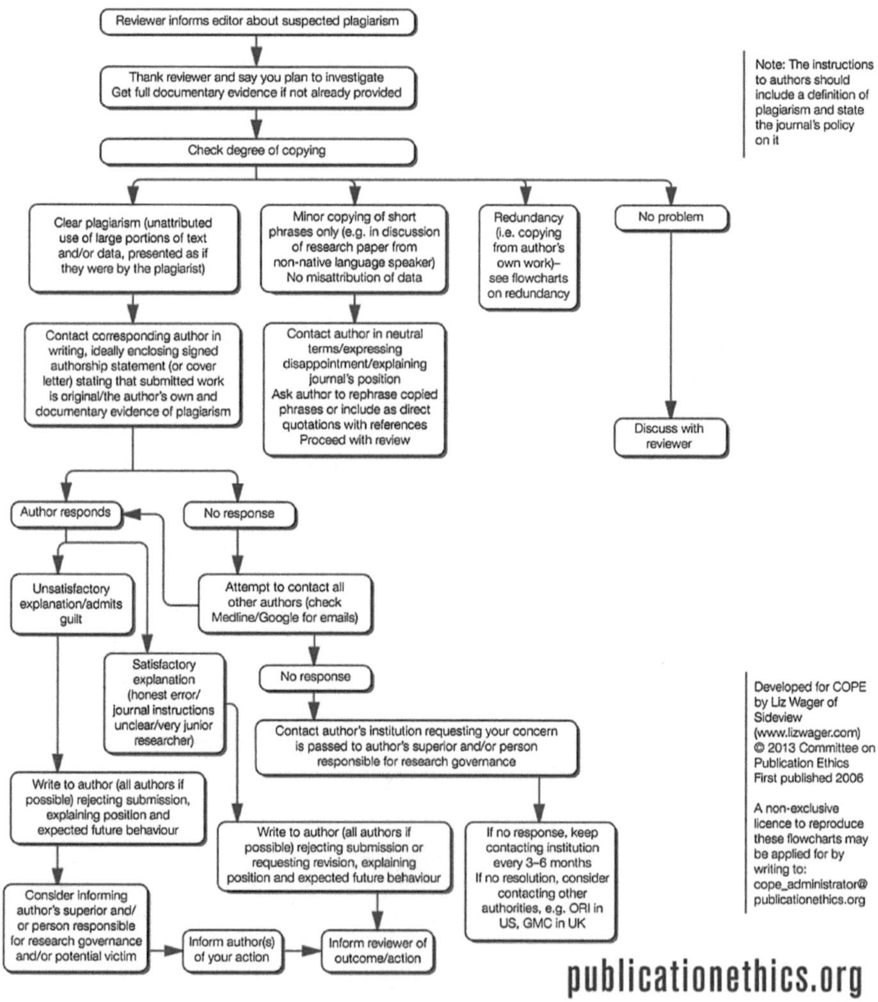

Fig. 10.1 a, b What to do if you suspect plagiarism in a submitted (**a**) and published (**b**) manuscript (COPE) (reprinted with the permission of COPE; http://publicationethics.org/resources/flowcharts)

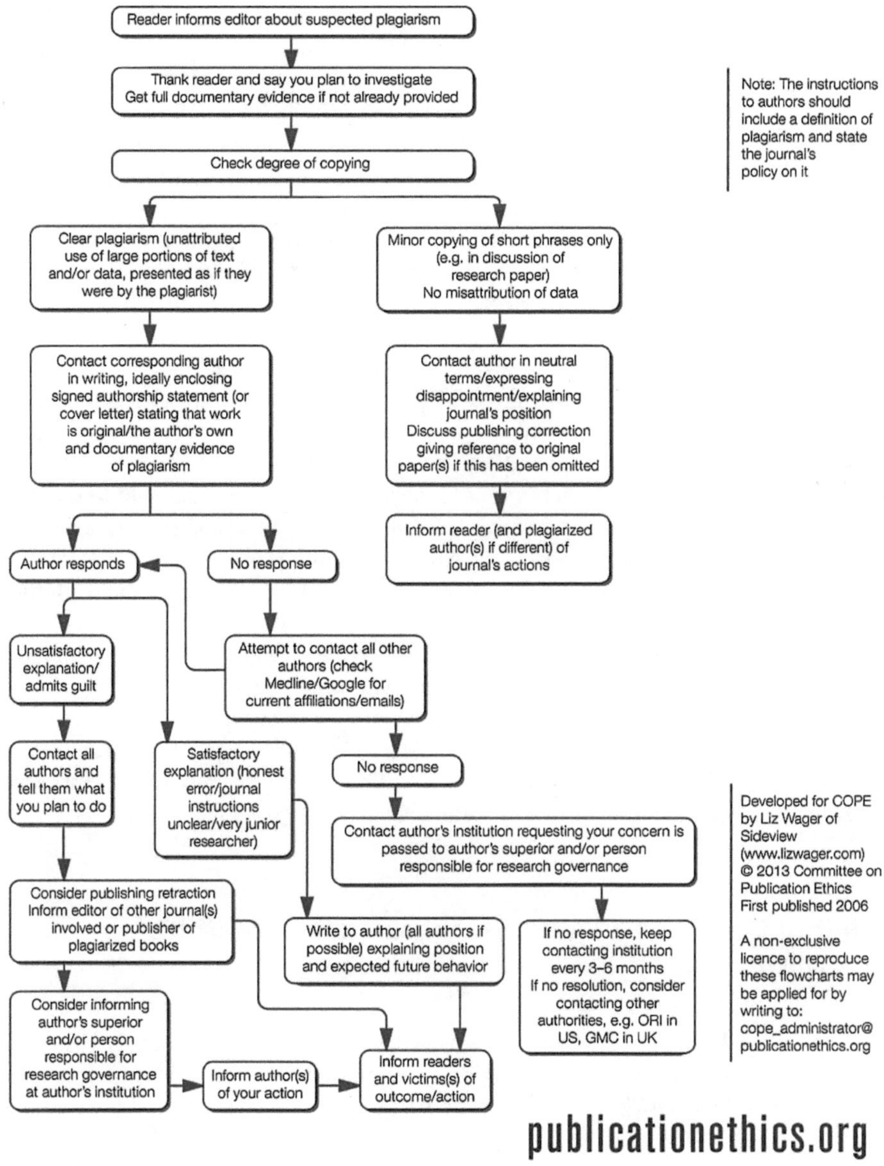

Fig. 10.1 (continued)

These charts are reproduced as Fig. 10.1a, b below (with the kind permission of COPE).

The COPE process outlined in Fig. 10.1a makes clear that the primary responsibility lies with the peer reviewer who initially raises the suspicion of plagiarism. However, this may be difficult if the reviewer has a heavy reviewing workload (say, more than 100–150 papers per year) and if many of these are from non-Anglophone countries. Of course, these days it is unnecessary to rely totally on reviewers to detect potential plagiarism. Tools such as CrossCheck (see Chap. 9) are able to automate the search for unexpectedly high levels of similarity with previously published texts. Nevertheless, CrossCheck cannot search non-textual items such as tables, charts and diagrams—and nor can it search for plagiarism of ideas. It is therefore sensible to ask reviewers to focus on these aspects.

At *JZUS* we have modified the COPE workflows slightly, as shown in Fig. 10.2a, b, to accommodate the special circumstances of our own journals: we have a very large number of submissions annually, and most of these are from non-English-speaking countries. As previously explained, we rely heavily on CrossCheck to highlight unusual levels of duplication in the textual content of submitted articles, since this is a good indicator of potential plagiarism. Our first step, unlike in the COPE workflow, is to run every submitted paper through CrossCheck (see Chap. 9 for details) in order to identify any textual similarities. With the similarity reports which are produced, the editors are then able to analyse potential plagiarism problems in more detail and identify the appropriate action to take. However, we are well aware that CrossCheck cannot itself prevent plagiarism. In the end, we are dependent on the hard work of our peer reviewers and journal editors to assure the academic integrity of the published materials.

The next step, shown in Fig. 10.1a, b (below), demonstrates the industry standard procedure for contacting an author or authors who may be suspected of plagiarism. COPE recommends a different approach depending on the type and degree of plagiarism, and on the author's response; this makes it clear just how much time and effort journal editors nowadays need to devote to handling this type of misconduct, in order to ensure academic integrity and correct attribution of copyright.

What to Do About Plagiarism When Detected—*JZUS'* Procedure

We use CrossCheck at two different stages in article processing. Figure 10.2a shows the first stage, when the article is originally submitted and before it is sent out for peer review. Figure 10.2b shows the second stage, after peer review and before final acceptance for publication.

First Check

The first thing our editors have to do is to examine the CrossCheck similarity reports: Overall Similarity Index (OSI) and Single Match Similarity Index (SMSI) [6]. If these reports show no significant overlap (duplication) between the submitted

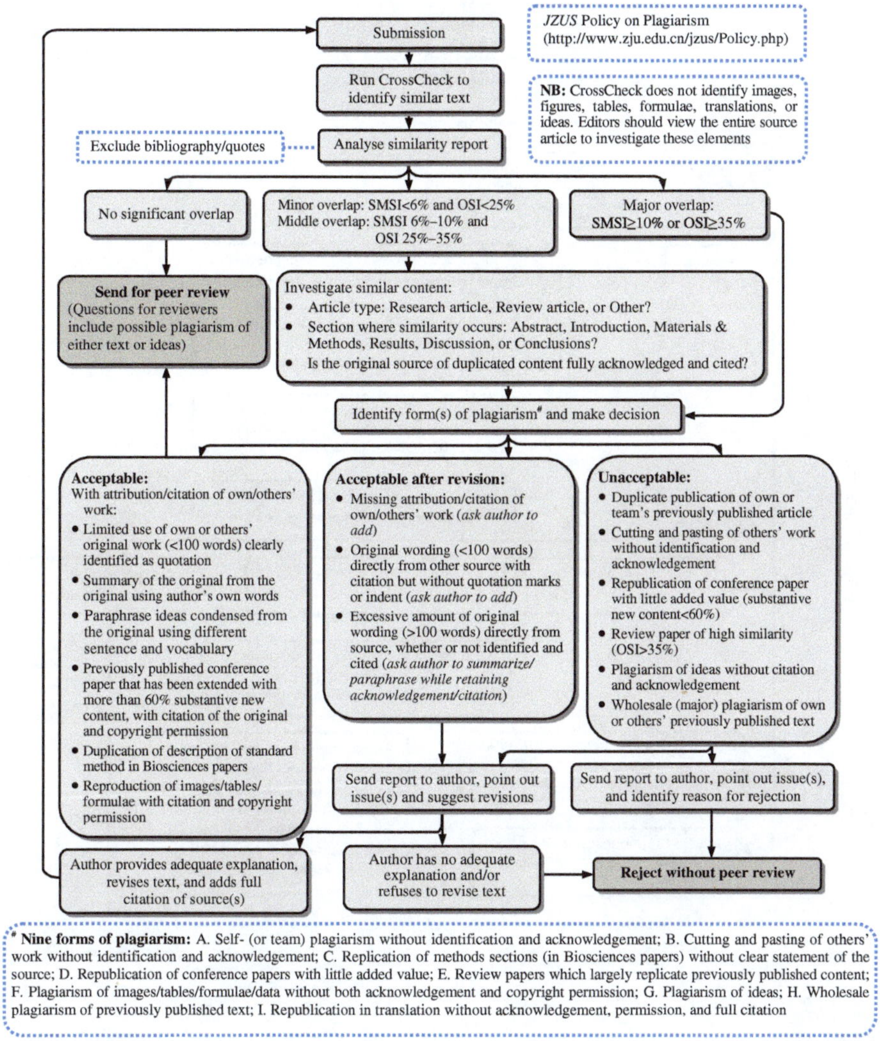

Fig. 10.2 a, **b** *JZUS*-Crosschecking Workflow during submission (**a**) and before publication (**b**) [*JZUS* makes it very clear to authors that, in addition to its high international peer review standards and its strict anti-plagiarism policy, CrossCheck will be used to check their submitted papers in order to identify unduly high levels of similar text, which gives rise to the possibility of plagiarism.]

article and other content in the CrossRef database (i.e. SMSI* < 6 %, OSI* < 25 %), then the article may be sent immediately for peer review.

However, if a significant overlap is indicated, the editors will need to examine the duplicated material in order to establish the nature of the problem. In our view,

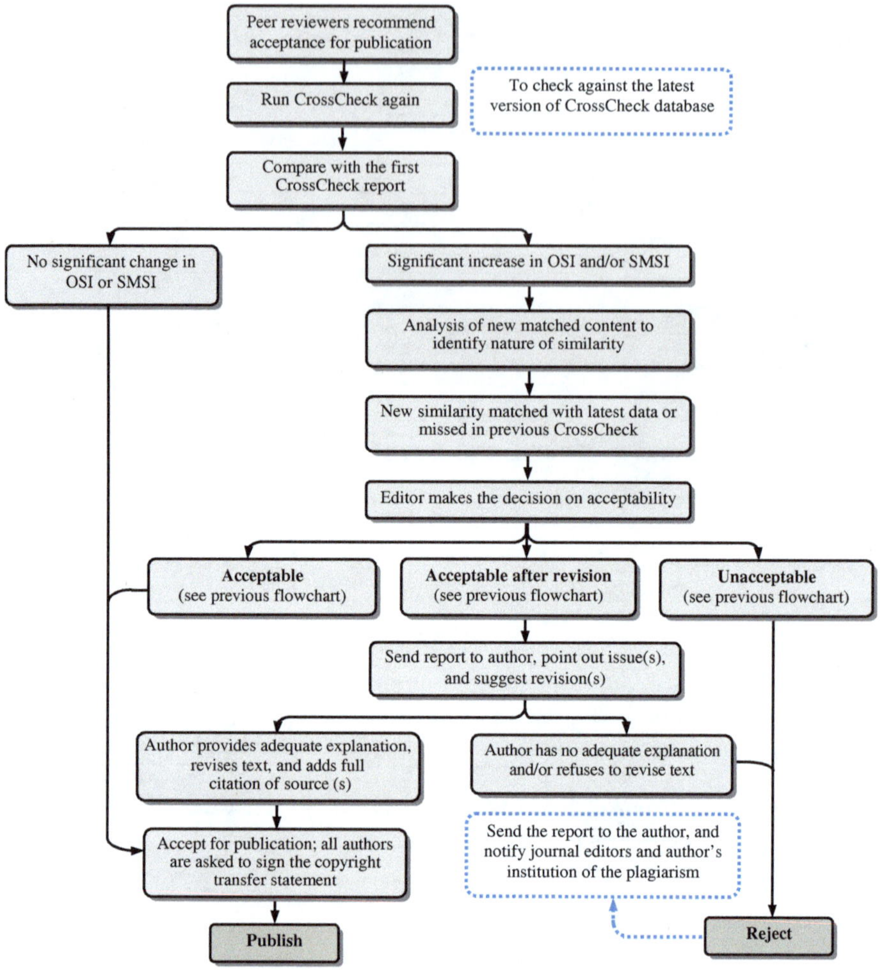

Fig. 10.2 (continued)

SMSI < 6 % and OSI < 25 % represents 'minor overlap'; in this case, our course of action would be as follows:

(1) If the duplicated material is acceptable in extent (i.e. less than 100 words [7, 8], or more with copyright permission. And it does not involve the significant content (merely only in small amounts to mimic sentences or words), and the previous publication is correctly attributed and cited, the paper would be sent for peer review;
(2) If the extent is acceptable, but the attribution/citation/permission is lacking, we would send the similarity report to the author, pointing out the issue(s) and suggesting revisions.

If the scores are higher—SMSI between 6 and 10 % and OSI between 26 and 35 %) we would consider this to be 'moderate overlap' may cause us alert, and our course of action would be as follows:

Investigate similar content:

(1) Article type: Research article, Review article, or Other?
(2) Section where similarity occurs: Abstract, Introduction, Materials and Methods, Results, Discussion, or Conclusions?
(3) Is the original source of duplicated content fully acknowledged and cited?
(4) Identify form(s) of plagiarism according to *JZUS*-policy and make decision (see Fig. 10.2a in detail)
(*OSI: the Overall Similarity Index represents the 'percentage of similarity between a submission and information existing in the iThenticate databases selected as search targets' (minor: <25 %; middle: 25–35 %; major: >35 %); *SMSI: the Single Match Similarity Index represents the percentage of similarity from a single source (minor: <6 %; middle: 6–10 %; major: >10 %)

If the SMSI is higher than 10 %, or the OSI is higher than 35 %, we would consider this to be 'major overlap', which is unacceptable in any circumstances. The editors would send the similarity report to author and reject the article immediately, directly without peer review.

Second Check

Our editors re-check each manuscript which the peer reviewers have recommended for publication, before final acceptance. This means that the paper will be checked against the latest version of the CrossCheck database (new material is being added to the database all the time). If the similarity reports, and the editors' own analysis of the results, suggest plagiarism, the editors will follow the same procedures as before. If major overlap is detected, our editors will also notify the editors of the journals in which the plagiarized material first appeared and—depending on the seriousness of the offence—they may also inform the author's institution. On average, we have detected about 2–3 cases of serious plagiarism per year at this late stage.

Reported Plagiarism

As mentioned above, the role of our expert peer reviewers is crucial in identifying plagiarized material which could not be detected by CrossCheck. Every year, we have a few cases where reviewers notify us that material has been published previously; we also occasionally receive information from other journals who run similar checks to our own, that they have discovered plagiarism of one of our own articles. For example, in 2014, a Japanese English-language journal discovered that large parts of a paper published in *JZUS* had already been published in their own journal; as a result, we took a very strong line with the authors, who immediately issued a retraction note (http://www.zju.edu.cn/jzus/article.php?doi=10.1631/jzus.B12r0102).

Does publication imply the end of the process? Obviously, the answer is 'no'. When a manuscript has been published, it is open to the scrutiny of every reader. If a reader (or the editor of another journal, as in the case above) informs our editors of plagiarism and provides the evidence, our editors will compare the provided evidence with the published manuscript to analyse the degree of overlap. If half or more of the contents are duplicated, we will pursue the matter with the authors with appropriate severity; if another journal is involved (whether plagiarizing or plagiarized), we will report the issue to the editors of the other journal.

Our investigation of suspected plagiarism, whether detected through CrossCheck or reported to us by a third party, is carried out according to the following principles:

1. If the problem relates to non-textual material (such as tables, figures or equations) we will carry out Internet searches to try to find the original sources. If it relates to text, we will forward the CrossCheck similarity reports. In either case, we contact the corresponding author of the article in question to seek an explanation.
2. If the overlap is 'minor' (see above), the author's response is positive and courteous, and he or she is willing to resolve the issue as suggested by the editor, processing can proceed.
3. On the other hand, if we have clear evidence of serious plagiarism, then we will reject the paper irrespective of the author's response; our aim in such instances is to raise authors' awareness of the importance of ethical publishing behaviour. In addition, when appropriate (for example, if author(s) have alleged serious plagiarism, in the fact of such irrefutable fact but still invent excuse and do not acknowledge the misconduct in order to educate authors and ensure the academic reputation) we would notify author's institution in order to further raise the profile of ethical issues.

Anti-plagiarism Policy of *Journals of Zhejiang University-SCIENCE*[1]

At *JZUS* we have identified nine basic forms of plagiarism that we typically encounter. These are:

- Self- (or team) plagiarism without identification and acknowledgement
- Cutting and pasting of others' work without identification and acknowledgement
- Replication of methods sections (in Biomedical journals) without clear statement of the source

[1](see Fig. 10.2a, b) (© *Journals of Zhejiang University-SCIENCE 2015*. This document was drafted by Zhang YH (Helen), and charted by Lin HF and Ye Q on May 11, 2015) [9].

- Republication of conference papers with little added value
- Review papers which largely replicate previously published content
- Plagiarism of images/tables/formulae/data without both acknowledgement and copyright permission
- Plagiarism of ideas
- Wholesale plagiarism of previously published text
- Republication in translation without acknowledgment, permission and full citation

A. **Self- (or team) plagiarism without identification and acknowledgement**

When an author (or another author from the same research project) repeats text and/or data from his/her (or the team's) previously published work, either verbatim or with minimal rewording, without clear acknowledgement, this is defined as self- (or team) plagiarism [3].

Remedy—identify the repeated (or reworded) content and provide full citation to the original publication.

B. **Cutting and pasting of others' work without identification and acknowledgement**

When an author directly copies text (or other content), either verbatim or with minimal rewording, from others' work, the quotation must be clearly identified (by quotation marks or—if necessary—indentation) and a full citation must be provided to the original source. Otherwise, it will be considered plagiarism [10].

Remedy—identify the quoted (or reworded) content and provide full citation to the original publication.

C. **Replication of methods sections (in Biomedical journals) without clear statement of the source**

When a standard method is identical to that described in a previously published article (whether by the same or different authors), it may be repeated verbatim, but a full citation must be provided to the original source; otherwise, the behavior will be considered plagiarism [8, 11].

Remedy—identify the quoted method and provide full citation to the original source. If the description is particularly long or complex, it may either be (a) appended to the paper as supplemental material or (b) provided in the form of a link to the published original, if this is freely available online [12].

D. **Republication of conference papers with little added value**

If the paper has not been substantially reworked to include additional detail which could not be included in the conference paper, such as detailed proofs or wider comparison with other work in the field, it would be considered self-plagiarism [13].

Remedy—add 60 % or more of substantive new material, which adds value to the original conference paper. Full citation to the original publication must be given, as well as copyright permission from the original publisher [13].

E. **Review papers which largely replicate previously published content**

If the summaries consist wholly or mainly of the original authors' words (OSI > 35 %), even if they are clearly identified with quotation marks, this would be considered plagiarism [6].

Remedy—the review author should rewrite the overview in his or her own words.

F. **Plagiarism of images/tables/formulae/data without both acknowledgement and copyright permission**

Reproducing illustrative content such as tables, diagrams, images or photographs, or indeed formulae, from someone else's work without both acknowledgement and (in the case of illustrative content) copyright permission from the original publishers, is considered plagiarism, and potentially also copyright infringement [14].

Remedy—provide full citation to the original source, and (in the case of illustrative material) obtain written copyright permission from the original publisher, and include an acknowledgement in whatever form they require.

G. **Plagiarism of ideas**

If an author reuses another author's ideas (the product of their intellectual effort) without acknowledgment of the original together with a full citation, this is considered plagiarism [10].

Remedy—identify the originator of the idea(s), and provide a full citation to the original published source.

H. **Wholesale plagiarism of previously published text**

If an author submits, as a new publication, large sections (or even complete articles) of his/her own or others' previously published text, we consider this to be major plagiarism and/or duplicate publication, and will always reject the article. An indicator of what constitutes 'large sections' would be if the SMSI is >10 % or the OSI is >35 %) [6].

Remedy—there is no remedy in this instance; the paper will always be rejected.

I. **Republication in translation without acknowledgment, permission and full citation**

If an author submits for publication an article which has already been published in another language without acknowledging the fact, this would be considered self-plagiarism. The fact that the article has already been published in another language must be clearly acknowledged and the original publication fully cited; copyright permission must also be obtained from the original publisher.

Remedy—make clear that it is a translation, provide full citation to the original publication, and obtain written copyright permission from the original publisher [15].

Sanctions

Authors who are found guilty of plagiarism [16–18] can pay a high personal price: they can lose their reputation, their position and even their academic awards. 'The personal punishments for plagiarizing can be much worse than the academic, workplace or legal punishment. It takes years to establish a reputation, and a single instance of plagiarism can tarnish a good name' [19].

It is not enough to provide policies, and to educate those who are unaware of the rules of acknowledgement, for cultural or other reasons (discussed in Chap. 1). There also need to be meaningful sanctions for those who are guilty of deliberate or repeated plagiarism. Nicholas Steneck, the head of the US Office of Research Integrity, stated, 'research fraud [including plagiarism] is not a victimless crime. Science is subverted, funds are wasted, and patients suffer and sometimes die as a direct result. So why is it not a real crime? [20] If we made it a criminal offence, subject to a country's criminal law, it would have to be taken seriously.' And in 2011, Wager raised a question, 'What sanctions should journals impose on authors for plagiarism?' [7], for which, The 1999 COPE 'Good Publication Practice' offers general guidance on the sanctions, as follows: (© 2011 COPE This is an open-access article distributed under the terms of the Creative Commons Attribution License, which permits unrestricted use, distribution, and reproduction in any medium, provided the original author and source are credited.)

'The following [sanctions] are ranked in approximate order of severity:

(1) A letter of explanation (and education) to the authors, where there appears to be a genuine misunderstanding of principles.
(2) A letter of reprimand and warning as to future conduct.
(3) A formal letter to the relevant head of institution or funding body.
(4) Publication of a notice of redundant publication or plagiarism.
(5) An editorial giving full details of the misconduct.
(6) Refusal to accept future submissions from the individual, unit, or institution responsible for the misconduct, for a stated period.
(7) Formal withdrawal or retraction of the paper from the scientific literature, informing other editors and the indexing authorities.
(8) Reporting the case to the General Medical Council, or other such authority or organisation which can investigate and act with due process [7]'.

Generally COPE recommends caution in applying sanctions against authors where there has been no due process of investigation, such as an institutional finding of misconduct. In fact, many universities and institutes have themselves introduced a clear sanctions policy; for example, Leiden University's website clearly states that 'Instances of fraud and plagiarism are taken very seriously by the Institute of Public Administration and will be subject to sanctions' [21].

Recently, *Nature*, asked ten experts in plagiarism (including the present author) about ways to prevent it, either among students or scientists. The responses were published in 2012 [22]. A number of proposals were made; for example, someone from universities suggested 'Teach scientists to paraphrase'; 'Check all manuscripts', or 'Catch system gamers'. The others from publishing area suggested 'Spot subtle forms'; 'Flag plagiarized'; 'Use professional translators'. Our own suggestion was that the current system of policing plagiarism is not sufficient. We proposed setting up an international database that 'Blacklists Frequent Offenders', which is more inclined to punitive proposal. For example, 'in many European countries, as well as in many US states and in China, a driver's licence comes with a point system. If you are caught breaking the law, by speeding, for example, you are issued penalty points. Too many points, and you lose your licence; getting it back is expensive and time-consuming'. Of course, 'the devil is in the details. Who would set up the database and monitor it? How many instances of plagiarism would be needed for someone to be blacklisted? All major publishers—commercial and nonprofit—should sign up to the project so they can work out the answers to such questions' [22]. Of course, we also agree with advice of Sandra Titus who is from US Office of Research Integrity, "Invest in prevention" is more important than buying software to detect plagiarism. Furthermore, 'Prevention efforts need to be directed to students, faculty members and institutions whose leaders must convey a consistent message on the importance of integrity.' However, we always believe, as same as the law of human society, it is necessary establishing a punishing system to curb academic misconduct.

In his recent discussion and based on his scientific experience worked in the Harvard-MIT Division of Health Science and Technology for more than 50 years and being some scientific journals' board members, professor Kiang not only in 22 years ago [23, 24] pointed 'that misconduct or plagiarism in scientific research and publishing areas are often not easy or quickly corrected. It is not only an ongoing subject but needs wider recognition if it is to be overcome or prevented', but his the latest email of June 6, 2015, emphasized 'simple copying of words or phrases is easily uncovered by computer programs but appropriation of ideas is impossible to even define, much less expose or punish. So, as a practical matter, plagiarism remains, for the most part, an intellectual offense, usually unpunished except for some damage to one's reputation. It is time to review our murky thoughts on plagiarism and its related broader concept of "Intellectual Property" on which there is a plethora of ideas in laws, ethics and philosophy.'

References

1. Butler, D. 2010. Journals step up plagiarism policing. *Nature* 466:167. http://www.nature.com/news/2010/100705/full/466167a.html.
2. Wager, E., and S. Kleinert 2011. Responsible research publication position statements chapter 49. In *Promoting Research Integrity in a Global Environment,* eds. T. Mayer., and N. Steneck, 305–307. Singapore: Imperial College Press/World Scientific Publishing. ISBN: 978-981-4340-97-7. http://publicationethics.org/files/International Stanards_Intro_for website_11 Nov 11.pdf.
3. Morris, S., et al. 2013. *Handbook of journal publishing (467 pp).* Cambridge: Cambridge University Press.
4. COPE: International standards for authors and editors. http://publicationethics.org/resources/international-standards.
5. COPE: Promoting integrity in research publication. http://publicationethics.org/.
6. Zhang, Y.H., and X.Y. Jia. 2012. A survey on the use of CrossCheck for detecting plagiarism in journal articles. *Learned Publishing* 25(4): 292–307. doi:10.1087/20120408.
7. Wager, L. 2011. How should editors respond to plagiarism? COPE discussion paper. Available from http://publicationethics.org/files/COPE_plagiarism_discdoc_26Apr11.pdf.
8. Roig, M. 2013. Avoiding plagiarism, self-plagiarism, and other questionable writing practices: a guide to ethical writing. The office of research integrity. Available from https://ori.hhs.gov/sites/default/files/plagiarism.pdf.
9. Zhang et al. 2015. Anti-plagiarism policy of *JZUS-A/B & FITEE*. *J Zhejiang Univ-Sci A (Appl Phys & Eng)* 16(6):507–512. http://www.zju.edu.cn/jzus/Policy.php.
10. University of Oxford. 2015. Plagiarism. Available from http://www.ox.ac.uk/students/academic/guidance/skills/plagiarism.
11. Jia, X.Y., X.F. Tan, and Y.H. Zhang. 2014. Replication of the methods section in biosciences papers: is it plagiarism? *Scientometrics* 98: 337–345. doi:10.1007/s11192-013-1033-5.

12. Zhang, Y.H., X.Y. Jia, H.F. Lin, and X.F. Tan. 2013. Be careful! Avoiding duplication: a case study. *Journal of Zhejiang University-SCIENCE B (Biomedicine and Biotechnology)* 14(4): 355–358. doi:10.1631/jzus.B1300078.
13. Zhang, Y.H., and X.Y. Jia. 2013. Republication of conference papers in journals? *Learned Publishing* 26: 189–196. doi:10.1087/20130307.
14. Saunders, J. 2010. Plagiarism and the law. *Learned Publishing* 23: 279–292. doi:10.1087/20100402]la.
15. Zhang, Y.H. 2013. Bilingual (multilingual) publications and duplicate publications for or against. *Journal of Zhejiang University-SCIENCE A (Applied Physics and Engineering)* 14(9): 687–690.
16. Pidd, H. 2011. German defence minister resigns in PhD plagiarism row. The Guardian, 1 March 2011. http://www.theguardian.com/world/2011/mar/01/german-defence-minister-resigns-plagiarism.
17. German education minister quits over PhD plagiarism. 2013. The Guardian, 9 February 2013. http://www.theguardian.com/world/2013/feb/09/german-education-minister-quits-phd-plagiarism.
18. A bold move against plagiarism. 2014. China Daily, 27 August 2014. http://usa.chinadaily.com.cn/opinion/2014-08/27/content_18495472.htm.
19. A story by Blue Angel. 2015. Rendering to the plagiarist(s): Five points arise. http://www.writerscafe.org/writing/_BlueAngel/1478685/.
20. Godlee F. 2007. Plagiarism and punishment. *BMJ*. 335(7627). doi:10.1136/bmj.39392.602523.47] http://www.ncbi.nlm.nih.gov/pmc/articles/PMC2071996/.
21. Fraud and plagiarism. http://campusdenhaag.leiden.edu/publicadministration/students/pre-masterstudents/regulations/plagiarism.html.
22. Comment: How to stop plagiarism. *Nature*, 481:22. doi:10.1038/481021a.
23. ORI/AAAS Conference on Plagiarism and Theft of Ideas in NIH. 1993. http://ori.hhs.gov/documents/aaas.pdf (DR. NELSON KIANG Speech 6-22-93).
24. Kiang, N. 1995. How are Scientific Correction Made? *Science and Engineering Ethics* 1: 347–356.

Chapter 11
Concluding Remarks: The Future of Plagiarism

As this book has made clear, plagiarism is an offence against research integrity itself. So, in looking forward, let us consider how the whole culture of research integrity might be improved.

Scientific Integrity Is of Global Concern

'Like other institutions, the institution of science has developed an elaborate system for allocating rewards to those who variously live up to its norms… The evolution of this system has been the work of centuries, and it will of course never be finished' [1]. This quote was highlighted on the website of the 4th World Conference on Research Integrity (4th WCRI, http://www.wcri2015.org) [2].

At the end of this book, before outlining the final concluding remarks, I want to say that two things happened recently, which are closely related to the book's theme of research integrity.

The first was in early April, as I was busy planning to finish the first draft at the end of May (deadline), when I received an invitation letter from Dr. Anderson, Co-Chair of the 4th WCRI that focuses on Research Rewards and Integrity. As a journal editor from China I felt very honored to give a talk at the plenary session of The Role of Publishers in Driving Change (http://www.wcri2015.org/speakers.html see Fig. 11.1 and Appendix 1). In this session, the first speech was from *Nature*'s editor, Dr. Kiemer, who asked 'What is the role of Journals and publishers in driving research standard changes?' and stressed six points for 'Role of Journals' on research integrity as follows:

Raise awareness (of research integrity);
Be a catalyst and facilitator of discussions;
Drive some changes;
Ensure full reporting, effective review and measured conclusions;
Provide opportunities for detailed and accurate credit for all contributions;
Respond quickly and thoroughly to criticisms of published papers.

Y.H. Zhang, *Against Plagiarism*, Qualitative and Quantitative Analysis of Scientific and Scholarly Communication,
DOI 10.1007/978-3-319-24160-9_11

Fig. 11.1 Confirmed plenary speaker: YH (Helen) Zhang reporting site in the 4th WCRI

Then, my presentation was 'Against Plagiarism: A global survey of Anglophones and non-Anglophones' and the final conclusion was:

The plagiarism detection tool and similarity report are very useful and effective, which can assist editors to screen documents suspected of plagiarism;
Global editors have expressed a strong mainstream view on ethical standards even though there are slight variations between different disciplines and countries, as well as between non-Anglophone editors and Anglophone editors;
A universal principle (policy) and practical approaches to prevent plagiarism and duplicate publications should be established.

In retrospect, I really appreciate the 4th WCRI's Committee giving me every opportunity to take part in the 4th World Conference on Research Integrity, flying from China to Brazil meaning from the Eastern to the Western Hemisphere, and then flying from the Northern to the Southern Hemisphere, in total a 36 h flight. Here it was exciting for me to feel a positive research integrity atmosphere in the world. And the conference brought together in total more than '500 participants from more than 50 countries' [3]. These policy makers, funders, research leaders, university and institutional leaders, scientists, students, country representatives, editors and publishers came together in order to discuss the state of research conduct and the research integrity system with the aim to learn from each other and debate possible solutions to strengthen the research system locally, nationally and globally for the benefit of society as a whole. Here each attendee would be mostly attracted by interesting topics, ideas, research results, discussions, recommendations, as well as some active critical proposals from different countries, departments, agencies, disciplines. And reviewing WCRI's background, we can know that since 2007 the WCRI has become an important ethics forum for scientists, academic publishers, research agencies, university education departments etc. For example, the 'Singapore Statement on Research Integrity'. (http://www.singaporestatement.org/)

and 'Montreal Statement on Research Integrity' (http://www.researchintegrity.org/Statements/Montreal%20Statement%20English.pdf) respectively launched at the 2nd and 3rd WCRI, which have become important guidance documents for research integrity in scientific and publishing areas.

In short, this experience not only made me learn much more about crediting research progress between scientific and publishing areas, but also realize that research integrity has become a global culture of science and publication.

The second thing that happened was when, on 6th June, I received an email (Appendix 2) from Professor Nelson Kiang who has worked in the Harvard-MIT Division of Health Science and Technology for more 50 years, and is our journal's editorial board member, I was thinking about how to end this book. My thanks to Prof. Kiang more active suggestions to my book and recommending his two papers [4, 5] published 22 years ago, which let me again know that misconduct or plagiarism in scientific research and publishing areas are often not easily or quickly corrected. It is not only an ongoing subject but needs wider recognition if it is to be overcome or prevented. Kiang pointed that plagiarism is the academic analogue of crime in the more general social context and can no more be eradicated than by expanding police enforcement. The problem is embedded in the reward structure of society, especially 'in the drive for "success" in the modern world', and 'how Institutionalized Science corrects errors from the viewpoint of a working scientist.' His view is more consistent with the 4WCRI's landmark quote—what Merton raised in 1957 [1].

From reality to history, we have seen scientific integrity is of global concern. So if we could change something for the institutionalization of science and publication in respect of scientific integrity, what is the first thing to do? Yes, it is time to promote a culture of greater transparency for scientific research and evaluation [6].

Transparency Is a Culture of Scientific Integrity

How to promote a culture of greater transparency?

I think we can take transparency as a magnifying glass of scientific evaluation.

We have to admit, in a sense, that the progress of academic research and publication is mainly based on previous literature [7]. However, both the public and the academic community need to know whether the research withstands verification and is truly new research, and who are the real contributors. At this point, in light of advances in network technology, and rampant plagiarism these days, it is time to call for an open and transparent policy to combat plagiarism. In other words, transparency is a prerequisite in scientific evaluation which will be like a magnifying glass looking at your research results and publications to say Yes (right, truthful, and original) or No (wrong, false and a copy).

In 1995, Prof. Kiang also stated 'I personally favor opening the scientific process to scrutiny so that all will have a more realistic view of how scientists work. On the other hand, there is nothing wrong in trying to live by the idealized creed taught to beginning students' [5].

In his keynote speech at the 4th World Conference on Research Integrity in 2015 [8], Dr. Lex Bouter firstly shared the top 12 misbehaviors ranked by 59 experts from research and publishing areas in the word; among them there are 5 items focused on plagiarism, they are:

Re-use content from your own paper without reference
Re-use previously published data without disclosure
Duplication publication without disclosure
Use published phrases or ideas of others without referencing
Use unpublished phrases or ideas of others without their permission.

And then, he made the point that greater transparency is urgently needed in the following aspects:

Identification of publication or report bias
Replication of data analysis with the same protocol or with a different design
Identification of other questionable research practices
Re-use of data for secondary analyses or pooled analyses.

Finally, Dr. Bouter urgently suggested that the research institutes need Open Data (including study protocol, data collection, an analysis plan, any conflict of interest, amendments etc.); the (journal) publishers need Open Access in order to meet the conditions of transparency (including proper acknowledgement; guarantees against breaches of privacy and misuse; Predefined study protocol for re-use of data etc.).

Recently in June 2015, COPE also released the latest version of Principles of Transparency and Best Practice in Scholarly Publishing [9] developed jointly by 4 scholarly organizations (COPE, DOAJ—The Directory of Open Access Journals, OASPA—The Open Access Scholarly Publishers Association, and WAME—The World Association of Medical Editors). It presented 16 criteria for Principles of Transparency include Copyright and Process for identification of and dealing with allegations of research misconduct, the latter of which indicates that 'publishers and editors shall take reasonable steps to identify and prevent the publication of papers where research misconduct has occurred, including plagiarism, citation manipulation, and data falsification/fabrication, among others…'.

As emphasized in the Project Proposal 'Ethics, Transparency and Integrity in Education' for the New Program of Activities 2016–2017 of the Council of Europe, we urgently need to develop a culture of transparency and integrity, specifically for universities, institutions, publishing areas [6]. Therefore, This is the illustration of 'Change the way we learn, study, assess and publish etc.'.

Honest Is the Best Policy for Research and Publication

In addition to transparency, honesty is crucial if one wants to succeed as an author, researcher, journal publisher, or editor. As Alisher Navoiy, the national poet of Uzbekistan, said as long ago as the 15th century:

> Truthfulness is the essence of honourable people.
> Two themes are seen within it.
> Firstly, be honest not only in words
> But also in thoughts and deeds.
> Secondly, scorn the world of falsehood
> But speak out the truth intrepidly.
> Both qualities are good in themselves
> And together mark greatness of spirit [10].

For scientific research and publication as Benjamin Franklin said 'honest is the best policy,' [11] indeed, honesty is the cornerstone of research and publication.

Responsibility Is the Foundation of Scientific Research and Publication

I would like to borrow Cicero's words [12] 'We are not born for ourselves, our country has given us the responsibility' to interpret the responsibilities of science, that is, researchers, authors, journal publishers and editors, we are not born for ourselves, our work has given us a real responsibility, that is to say, we must be responsible for the facts and truth.

In 19 May, 2015, the Japan Society for the Promotion of Science Editing Committee (JSPSEC) just issued a book, entitled 'For the Sound Development of Science—The Attitude of a Conscientious Scientist' [13]. The first section stated 'what is responsible research activity' and described science as being is built upon a foundation of trust. Scientists have to be responsible for gathering data carefully, using appropriate analytical and statistical techniques, reading more previous writings or printing information and finally reporting their results accurately.

As previous chapters have described, over recent years, it is good news that many institutions, universities, publishers and journals have posted online their policies or statements about ethics and the responsibilities of researchers and authors.

In particular, since ORCID (Open Researcher Contributor Identity) was launched in 2012, it is basically for solving the problem of identifying the authors contributions and their responsibility in a transparent manner because ORCID has the ability to reach across disciplines, research sectors and national boundaries and is 'a hub that connects researchers and research through the embedding of ORCID identifiers in key workflows, such as research profile maintenance, manuscript submissions, grant applications, and patent applications' [14].

Sanctions Is Necessary in Scientific and Publishing Areas

As already described in Chap. 10, some researchers or authors have found guilty of plagiarism can pay a high personal price due to their academic misbehavior. However, during discussion with Prof. Kiang about 'Thoughts on Intellectual Property and Practical Implications' (see Appendix 2), he mentioned 'for the most part, an intellectual offense is usually unpunished except for some damage to one's reputation.' So today as a senior scientist, he strongly stresses it is time to think of 'enforcing edicts against plagiarism; perhaps an analogy may be made to the history of enforcing laws against drugs and alcohol etc.'.

Personally, I think in research and publishing areas as well as in other areas, if we do not punish errors (misbehavior) this would be tantamount to accepting the misbehavior, so in 2012, we proposed setting up an international database that 'Blacklists Frequent Offenders', which is more inclined to be a punitive proposal [15, 16]. We hope all major publishers—commercial and nonprofit making—would sign up to the project of an international database Blacklisted Frequent Offenders, so we can work jointly to take it as a solution to stop plagiarism. Certainly we know that only a blacklist alone will not stop plagiarism. Punishment alone will not stop plagiarism. The culture that accepts plagiarism has to be changed and it will take a long time. This is a moral issue, like the abolition of slavery in the USA in the 19th century. Laws by themselves did not stop slavery. Only the acceptance that this was morally wrong finally stopped it.

To sum up, Integrity, Transparency, Honesty, Responsibility and Sanctions, these are the key words in the fight against plagiarism in the past, present, future.

Concluding Remarks

According to the above, we can predict that the Future of Plagiarism almost has a limited "market" because these strict policies and new digital techniques (ORCID, CrossCheck etc.) are the bane of it, although we have to admit that plagiarism is an eternal topic in human society, and we cannot stop all plagiarism, which will always exist somewhere, just as there will always be drugs, always corruption etc.

More police cannot stop it. But we can positive steps to reduce it, don't you think?

Below are my six concluding points:

1. The whole world should pay attention to research integrity
2. Create a culture of transparency in science and publication
3. Take the honesty as the best policy for researchers and authors
4. Make responsibility the foundation of scientific research and publication
5. Sanctions are necessary in scientific and publishing areas
6. We can never completely eradicate plagiarism, but we have to make it morally and culturally unacceptable!

References

1. Merton, R. K. 1957. The reward system of science. In *On social structure and science*, ed. P. Sztompka, 286–304. Chicago: University of Chicago Press, 1996. http://www.wcri2015.org (This quote was highlighted on the website of the 4th World Conference on Research Integrity).
2. The 4th World Conference on Research Integrity. Research reward and integrity-improving systems to promote responsible research. http://www.wcri2015.org/.
3. Welcome from the conference co-chairs. http://www.wcri2015.org/4WCRI_Program.pdf May 31, 2015.
4. ORI/AAAS Conference on Plagiarism and Theft of Ideas in NIH. 1993. http://ori.hhs.gov/documents/aaas.pdf (DR. NELSON KIANG Speech 6-22-93).
5. Kiang, N. 1995. How are scientific correction made? *Science and Engineering Ethics* 1: 347–356.
6. International Conference Plagiarism across Europe and Beyond 2015, June 10th–12th, 2015 https://plagiarism.pefka.mendelu.cz/?sec=cp.
7. Collins, F. S., and L. A. Tabak. 2014. NIH plans to enhance reproducibility. *Nature* 505 (7485): 612–13. http://www.ncbi.nlm.nih.gov/pmc/articles/PMC4058759/.
8. Bouter, L. 2015. What is holding back us in the prevention of questionable research practices? Keynote in 4WCRI. http://www.wcri2015.org/speakers.html.
9. COPE. Principles of transparency and best practice in scholarly publishing. http://publicationethics.org/resources/guidelines-new/principles-transparency-and-best-practice-scholarly-publishing.
10. Navoiy, A. http://sambuh.com/en/uzbekistan/cities/tashkent/alisher-navoiy.html.
11. Benjamin Franklin quotes. http://www.brainyquote.com/quotes/quotes/b/benjaminfr151625.html.
12. Cicero Quotes (607 quotes). http://izquotes.com/quote/282137.
13. JSPSEC. 2015. For the sound development of science—The attitude of a conscientious scientist, 157. Maruzen Publishing.
14. ORCID. http://orcid.org/.
15. A story by Blue Angel. 2015. Rendering to The Plagiarist(s): Five Points Arise. http://www.writerscafe.org/writing/_BlueAngel/1478685/.
16. Comment: How to stop plagiarism. *Nature*, 481: 22. doi:10.1038/481021a

Appendix 1
An Invitation Letter from Dr. Anderson, Co-Chair of the 4th WCRI

Dear Professor Yuehong Zhang,

On behalf of the Planning Committee, it is my great honor to invite you to deliver a speech at the plenary session at the 4th World Conference on Research Integrity (www.wcri2015.org), to be convened in Rio de Janeiro, Brazil, from May 31 to June 3, 2015. The World Conference on Research Integrity is the primary global forum to address research integrity. The focus of the conference is on improving systems to promote responsible research. It will bring together top leaders worldwide who have responsibility for research integrity and the responsible conduct of research. Speakers and participants – including leaders of professional societies, research funders, university presidents, journal editors, policy makers, and researchers – will discuss strategies for ensuring research integrity and extending global harmonization of standards and practices. Previous conferences have brought together participants from over 50 countries, and this diversity is already expected for this forthcoming event.

We thus look forward to welcoming you for the 4th WCRI in Rio de Janeiro, Brazil.

Kind regards,

Melissa S. Anderson, PhD
Co-Chair, 4th World Conference on Research Integrity

Y.H. Zhang, *Against Plagiarism*, Qualitative and Quantitative Analysis of Scientific and Scholarly Communication,
DOI 10.1007/978-3-319-24160-9

Appendix 2
Letter (Email) from Professor Nelson Kiang

Re: "Thoughts on Intellectual Property and Practical Implications"

2015-07-06 22:25:28

Dear Helen:

Thank you for sending me the Guidelines for the Anti-Plagiarism Polity of the Journal of Zhejiang University. You are doing good work.

Plagiarism and the broader topic of academic and research misconduct has long been one of my side-interests. I used to teach "Real-World Ethics" at MIT and "Research Practices" in a joint Harvard-MIT program in Speech and Hearing, along with my scientific work.

I rooted out a couple of papers I wrote when I was more actively participating in what has now become a world-wide issue. Perhaps you might be interested in reading about my personal experiences. As you can see, plagiarism is the academic analogue of crime in the more general social context and can no more be eradicated than by expanding police enforcement. The problem is embedded in the reward structure of society, especially in the drive for "success" in the modern world.

…………

2015-07-09 22:50:08

I gather that you might welcome suggestions on how to close your forthcoming book on plagiarism. Perhaps I can offer you some of my thoughts. One way to close your book could be with a chapter on "The Future of Plagiarism."

My American Eleventh Edition Merriam-Webster's Collegiate Dictionary defines "plagiarize" as "to steal and pass off (the ideas or words of another) as one's own: use (another's production) without crediting the source)". The British Concise Oxford Dictionary defines "plagiarize" as "take and use (the thoughts, writings, inventions etc. of another person) as one's own." The broadness of these typical definitions reveals why the act is not usually considered a criminal offense. There is simply no way to assure enforcement or even to determine applicability of general definitions to specific instances.

How to deal with plagiarism is only one example of a concept that needs to be openly debated by all involved peoples, with perhaps room for many different local

Y.H. Zhang, *Against Plagiarism*, Qualitative and Quantitative Analysis
of Scientific and Scholarly Communication,
DOI 10.1007/978-3-319-24160-9

resolutions. There may be no satisfactory policy that would cover all circumstances. Accordingly, schools and publishers can define their own policies and sanctions. Simple copying of words or phrases is easily uncovered by computer programs but appropriation of ideas is impossible to even define, much less expose or punish. Only the most naïve would copy whole papers or sections. It is easy to paraphrase or alter a few words. (I personally know of Chinese “scholars” who simply translate documents and pass them off as their original writings).

As to assigning credit for ideas (which often has monetary value or professional status consequences) disputes can, to say the least, be controversial and remedies unclear. So, as a practical matter, plagiarism remains, for the most part, an intellectual offense, usually unpunished except for some damage to one’s reputation. It is time to review our murky thoughts on plagiarism and its related broader concept of “Intellectual Property” on which there is a plethora of ideas in laws, ethics and philosophy.

There is an old saying that “There is nothing new in the universe”, as far as human thinking is concerned. Thus the very concept of original ideas (if not words) is undercut and newness is restricted to old ideas in new contexts. In one sense, all our experiences and familiarity with the thoughts of others determine the development of our individual views. A wry commentary on academic endeavors is “stealing from one source is plagiarism, stealing from many is scholarship”. In technology, reverse engineering is a widespread, even accepted, means of advancing social goals.

At the broadest level, the ease of distributing ideas and knowledge must surely be beneficial to society as well as individuals. Currently there is some debate as to whether patents and copyrights, once meant to incentivize innovation by economic means, now stifles innovation. (Indeed there are companies that buy up undeveloped patents simply to extract profits without playing any role in initiating ideas).

In thinking about enforcing edicts against plagiarism, perhaps an analogy may be made to the history of enforcing laws against prostitution, drugs and alcohol, sexual behaviors, pornography, profanities, irresponsible profiteering, insider trading, etc. The lesson has been simple: enforcement is spotty and rarely effective in eradicating all such bad behavior. Police, censors, and the like are overwhelmed and frustrated whenever highly publicized “wars” on crime, corruption, or other unethical practices are publically trumpeted. In fact, it can be argued that innovation is maximized when rules are minimized. Chinese history tells us that the chaotic periods between dynasties are exactly when intellectual ferment is most evident. New ideas emerge more easily during disorder than during periods of stability when incremental progress is more common. We could all profit from a world-wide open discussion of intellectual property at the broadest possible level.

On balance, the Yin and Yang of human thought exemplifies the tension between social order and creative freedom. The diversity of cultures, genes, and geographies is not easily suppressed nor perhaps desirable to discourage. Why not review how people think throughout the world and over history?

These are only some of my ideas, briefly jotted down hurriedly with the hope that they might help your thinking about this very important issue.

Sincerely yours,
Nelson Yuan-sheng Kiang
Eaton-Peabody Professor (Emeritus)
Harvard-MIT Division of Health Sciences and Technology and Department of Brain and Cognitive Sciences
Massachusetts Institute of Technology

Professor of Physiology (Emeritus)
Department of Otology and Laryngology
Harvard Medical School

Index

Y.H. Zhang, *Against Plagiarism*, Qualitative and Quantitative Analysis of Scientific and Scholarly Communication,
DOI 10.1007/978-3-319-24160-9

FSC
www.fsc.org
MIX
Papier | Fördert
gute Waldnutzung
FSC® C083411

Kolibri 360 GmbH
Ferdinand-Jühlke-Straße 7
99095 Erfurt, Deutschland
produktsicherheit@kolibri360.de